D0914318

MICROPROCESSOR-BASED PARALLEL ARCHITECTURE

for Reliable Digital Signal Processing Systems

CRC Press
Computer Engineering Series

Series Editor
Udo W. Pooch
Texas A&M University

Published books:

Telecommunications and Networking
Udo W. Pooch, Texas A&M University
Denis P. Machuel, Telecommunications Consultant
John T. McCahn, Networking Consultant

Spicey Circuits: Elements of Computer-Aided Circuit Analysis
Rahul Chattergy, University of Hawaii

Microprocessor-Based Parallel Architecture for Reliable Digital Signal Processing Systems
Alan D. George, Florida State University
Lois Wright Hawkes, Florida State University

Forthcoming books:

Discrete Event Simulation: A Practical Approach
Udo W. Pooch, Texas A&M University
James A. Wall, Simulation Consultant

Algorithms for Computer Arithmetic
Alan Parker, Georgia Institute of Technology

Handbook of Software Engineering
Udo W. Pooch, Texas A&M University

MICROPROCESSOR-BASED PARALLEL ARCHITECTURE for Reliable Digital Signal Processing Systems

Alan D. George, Ph.D.
Electrical Engineering Department
Florida State University and Florida A&M University
Tallahassee, Florida

Lois Wright Hawkes, Ph.D.
Computer Science Department
Florida State University
Tallahassee, Florida

CRC Press
Boca Raton Ann Arbor London Tokyo

Library of Congress Cataloging-in-Publication Data

Catalog record is available from the Library of Congress.

Direct all inquiries to CRC Press, Inc., 2000 Corporate Blvd., N. W., Boca Raton, Florida, 33431.

International Standard Book Number 0-8493-7176-7

Printed in the United States 1 2 3 4 5 6 7 8 9 0

Printed on acid-free paper

PREFACE

Major advancements have been made in the areas of parallel and distributed computer architectures, digital signal processing (DSP) microprocessors, and fault-tolerant computing. While each discipline continues to grow and develop, little has been done to combine their features. By doing so, a number of improvements in the DSP area can be made using microprocessor-based distributed multiprocessor systems.

This text addresses two such important improvements. The first relates to the need for DSP systems with higher levels of both performance and reliability. The second addresses the need for methods to design and develop new tightly-coupled multiprocessor systems and distributed multiprocessor systems in a way that reduces the escalating costs of this extremely complex task.

In addressing these two improvements, a new microprocessor-based parallel architecture and system is presented which provides much higher levels of performance than traditional single-processor DSP designs. In addition, the proposed design also increases system reliability by tolerating single faults. A new system prototype is developed without the prohibitive costs normally associated with such designs. This prototype is implemented completely in software, employing state-of-the-art simulation techniques to accurately forecast system response in terms of both performance and fault tolerance. The system prototype is studied and analyzed using such DSP applications as digital filtering and fast Fourier transforms.

Intended readers for this text generally include professionals in the fields of digital signal processing, parallel and distributed computer architecture, or fault-tolerant computing. However, engineering and computer science students and practitioners of all types will find this information useful, as sufficient background information is provided for those who desire to learn more in these areas.

The first chapter provides an introduction to the entire text. Topics include the problem statement in terms of both performance and fault tolerance, as well as the method used to address these problems.

The second chapter is a background overview of fault-tolerant computing, including such issues as basic definitions, redundancy techniques, communication architectures, properties, and examples of previously designed fault-tolerant systems. The third chapter is an overview of parallel computing, including such issues as parallel processing overview and definitions, parallel architecture models, operating system overview, and real-time computing overview.

In the fourth chapter, an overview of digital signal processing, basic algorithms, DSP microprocessors, and the Motorola DSP96002 device is presented. These issues, along with those in two previous chapters, are then used to discuss and develop the implementation requirements for basic DSP operations, as well as the most effective architecture to address these requirements.

In the fifth chapter, a set of premises and goals are identified for the system, and these parameters are used as the basis for the design and development of a new fault-tolerant computing system. Topics include such considerations as DSP implementations, redundancy techniques, communication architectures, clock synchronization, communication primitives, interface and communication design, and system design enhancements.

The sixth chapter describes how highly accurate simulation is used to essentially develop a prototype for this system entirely in software. This simulation provides an exceptionally accurate description of system behavior. The simulation is used as the basis for the seventh chapter, which describes and quantifies a number of system tests and evaluations. This chapter also includes a discussion on reliability modeling which provides a more analytical description of the fault-tolerant capabilities of this system.

The eighth and final chapter provides the conclusions drawn from this research. Issues addressed include such considerations as system performance, fault tolerance, and growth potential. An overall research summary is included, with consideration of such key points as: problems or questions addressed by this research; objectives; approaches used to achieve these objectives; work performed; and important results and conclusions.

Finally, a set of appendices is included with listings of the software developed for this text. This includes C code for multiprocessor simulation and DSP96002 assembly-language code executed on the processors themselves.

The authors would like to express their gratitude to all those who made this book possible. We thank all our colleagues who provided valuable suggestions and comments for revising this manuscript, both in its early form and later drafts, including Fred O. Simons, Jr. and Gideon Langholz of the FAMU/FSU College of Engineering, Abe Kandel of the University of South Florida, and David Kopriva of the Florida State University. In addition, we would like to thank Jack and Ruby Booker for providing the computer system upon which most of this research was conducted, as well as Janice Morey, Russ Hall, Joel Claypool, and the reviewers of CRC Press, Inc., for all their help in bringing this book to publication.

Finally and most importantly, we would like to thank our spouses and children for their constant love, patience, and support during the long period of time it took to perform this research and write this book. In particular, many thanks are due to the George family (Elizabeth, Andrew, and Jennifer) and the Hawkes family (Fred, Lara, Elliot, and Graham).

Alan D. George
Lois Wright Hawkes
Tallahassee, Florida

About the Authors

Alan D. George received the B.S. degree in computer science and the M.S. degree in computer engineering from the University of Central Florida, Orlando, Florida, in 1982 and 1985 respectively, and the Ph.D. degree in computer science from the Florida State University, Tallahassee, Florida, in 1991. He is presently Assistant Professor in the Electrical Engineering Department of the FAMU/FSU College of Engineering, a joint college of both the Florida A&M University and the Florida State University. Previously he was task leader and senior computer engineer for Martin Marietta, Orlando, Florida, computer engineer for General Electric, Daytona Beach, Florida, and programmer/analyst for the University of Central Florida at the Naval Training Systems Center, Orlando, Florida. His research interests are in parallel and distributed computer systems, fault-tolerant computer systems, microprocessor-based systems, and digital signal processing implementations.

Dr. George is a member of Tau Beta Pi, Upsilon Pi Epsilon, IEEE, ACM, and the American Society for Engineering Education.

Lois Wright Hawkes received the B.S. degree in pure mathematics from the University of Western Ontario, London, Canada, the M.S. degree in computer science from the University of British Columbia, Vancouver, Canada, and the Ph.D. in electrical engineering from the Imperial College of Science and Technology, University of London, London, England, in 1972, 1974, and 1977, respectively. She is presently Associate Professor in the Computer Science Department of the Florida State University, Tallahassee, Florida, where she has been a member of the faculty since 1981. Previously she was on the faculty in the Computer Science Department, School of Advanced Technology, SUNY-Binghamton, Binghamton, New York, and senior engineer for Sperry-Rand at NASA, Langley, Virginia. Her research interests include interconnection networks, fault tolerance, combinatorial filing schemes, intelligent tutoring systems, and distributed computing.

Dr. Hawkes is a member of the IEEE, ACM, the American Association of Artificial Intelligence, and Computer Scientists for Social Responsibility.

TABLE OF CONTENTS

Dedicated to my wife Elizabeth
whose love is my foundation ADG

To my parents
George and Myrtle Wright LWH

Chapter 1
INTRODUCTION

Digital Signal Processing (DSP) is based on the representation of signals by sequences of numbers and the processing of these sequences. The basic purpose of DSP is to estimate characteristics of a signal or to transform a signal into a more desirable form. As recent developments have illustrated, more signal processing implementations are using DSP techniques than ever before.

A fault-tolerant system is one that is capable of correct operation despite the presence of certain hardware failures and software errors. By applying fault-tolerant techniques in the design of a new computer system, the system can be made significantly more reliable.

A number of improvements are needed in the field of DSP systems, including: increased reliability, performance, and growth potential; and improved simulation methods and techniques for test and evaluation. However, in order to begin to achieve these goals, a number of critical developments have been lacking until recently. These include high-performance IEEE-754 floating-point DSP microprocessors, multiple-device simulation software callable from a high-level language, and simulation software capable of single-cycle granularity and complete device pinout support.

Based on these developments, the potential now exists to design a new fault-tolerant DSP computing system for a broad class of DSP applications. This design emphasizes fault tolerance and parallel processing and is prototyped completely in software. This in turn leads to system test and evaluation for a fraction of the traditional cost, and makes more advanced studies possible such as fault injection and analysis.

1.1 Statement of the Problem

The areas of fault-tolerant computing and high-performance DSP microprocessors have, to some extent, developed independently. In the past, the use of fault-tolerant techniques and redundant processors has been limited to the most critical of applications, primarily due to the cost of processors and their limited performance. But, as computer technology and applications continue to develop, the need for reliable high-performance computers is becoming evident in an ever growing number of applications. This is especially true with DSP systems. The increasing role of DSP computing systems has implied a greater need for more fault-tolerant systems with improved performance.

1.1.1 Fault tolerance

Some of the goals of fault tolerance are reliability, availability, safety, performability, maintainability, testability, and dependability [JOHN89a]. Traditional applications of fault-tolerant computing have included long-life applications, critical-computation applications, maintenance postponement applications, and high-availability applications. However, the need for some or all of these fault-tolerance goals in modern DSP computing systems is becoming more desirable. For example, as DSP microprocessor chip designs begin to exceed 1 million transistors per chip, the ability to test and verify correct operation becomes more difficult to achieve. By incorporating a fault-tolerant design in the overall computing system itself, of which these chips would be a part, many of these problems can be anticipated and overcome.

A number of factors must be considered in the design of a fault-tolerant computing system. Some of these are the interconnection network, the operating system, the application software, and the processors themselves. Some of the techniques for fault-tolerant design include passive hardware redundancy, active hardware redundancy, hybrid hardware redundancy, information redundancy, time redundancy, and software redundancy. And, to a great extent, the need for system as well as device fault tolerance and testability must be considered from the beginning. That is, in order to achieve the fault-tolerance goals desired, fault-tolerant techniques must be applied from the very beginning of the design process, and not merely as a retrofit. These and other concepts will be defined and considered in chapter 2.

For example, one major consideration in fault-tolerant distributed and multiprocessor computing systems is the communication architecture to be used. Pradhan describes seven fault-tolerant communication architectures [PRAD86a]. Each approach has associated with it a number of advantages and disadvantages which must be weighed with respect to the potential application (i.e. DSP) and the fault-tolerance goals desired.

1.1.2 Performance

Many believe that the 1990s will be the decade of distributed computing whereby computations will be divided into sub-computations that can be executed in parallel on multiple processors in a single computer system [DEIT90]. While the performance of single-processor systems continues to increase, their rate of growth is insufficient to keep up with many scientific computing needs. In fact, many experts believe that the gains being achieved with new uniprocessor systems will begin to level off as the limitations of single-processor technology begin to appear. A parallel processing approach will be required to achieve major improvements in execution speed, throughput, reliability, and availability. Whether accomplished via tightly-coupled multiprocessors using shared memory, or distributed loosely-coupled multiprocessors where each processor uses its own local memory, it is widely accepted that multiprocessing represents a major growth path for the future.

Modern DSP microprocessors such as the Motorola DSP96002 provide computational capabilities never before possible on a single chip. By combining *multiple* devices, processing speeds can be obtained which rival modern supercomputers. However, while it is currently not unusual to use DSP microprocessors in a relatively simple tightly-coupled shared-bus multiprocessor architecture, these efforts do not address the need for fault-tolerant machines. In fact, many of the techniques used to date (e.g. a simple shared-bus architecture) can be among the least effective approaches from a fault tolerance standpoint.

Another important problem is that of designing, implementing, testing, and evaluating new multiprocessor and distributed computing systems. In recent years, the cost of developing these systems has experienced exponential growth, often requiring thousands of man-hours and millions of dollars. Techniques need to be applied to reduce these growing costs.

In order to realize a new fault-tolerant computing system, it is not sufficient to simply perform a paper design and evaluate it. Experience has shown that many problems can arise when dealing with multiprocessor and distributed systems that are not recognized at the preliminary design stage. As noted by Tanenbaum, a leading expert in distributed operating systems, "totally unexpected problems ... make it necessary to build and observe real [distributed computing] systems to gain insight into the problems. Abstract formulations and simulations are not enough." [TANE85]

However, developing complex systems with multiple processors has continued to be a difficult, time consuming, and expensive task. Designing and implementing a prototype multiprocessor computer can often involve thousands of man-hours and hundreds of thousands of dollars. This limitation is no longer necessary with current software technology. By using DSP microprocessor simulation software (like that offered by Motorola for the DSP96002) as a starting point, it is now, for the first time, possible to design, implement, and evaluate a multiprocessor completely in software. This software enables multiple processor simulation to take place in parallel, with single clock-cycle granularity. By writing appropriate interface code in C, almost any architecture can be simulated. And, although not executing in real-time, the simulation can keep an accurate count of clock cycles executed, thereby allowing easy scaling to actual execution times. *As opposed to analytical models and studies, these simulations will exactly parallel the real ma-*

chines, without having to actually build them in hardware. In fact, the results are in effect prototype machines built in software.

As noted by Lee, the performance requirements of a continually increasing number of applications dictate the need for parallel processing in the form of multiple cooperating DSP microprocessors [LEE88, LEE89]. However, only minimal support has been available for designing systems with multiple DSP microprocessors until recently. Some of the newer DSP devices provide controllable wait states for access to external memory subsystems which could be used to manage the delays associated with contention in a shared memory system. Other devices provide special hardware pins and instructions for the synchronization of several tightly-coupled processors. Still others even provide duplicated expansion ports for interprocessor communication and high-bandwidth memory access.

Even these enhancements provide only minimal support for designing multiple DSP microprocessor systems. A more critical requirement that has not been addressed is the need for software simulators which support these multiple device systems. Historically, system designers have had to build their systems first and then test them, which often might imply constant re-designs until the tests pass successfully. One notable exception to this trend has been Motorola and their DSP56001 DSP fixed-point microprocessor. The Motorola software support includes a simulator which comes in the form of a library of functions or subroutines that are callable from user code. Each call to one of these simulation functions simulates the effects of a state change in a particular DSP56001 microprocessor for one clock cycle. By writing software in C to simulate the system interconnection network, shared memory subsystem, and any other hardware above and beyond the processors themselves, a system designer can effectively build a multiple processor prototype in software.

More recently, Motorola has announced the DSP96002 floating-point chip and has released a DSP96002 version of the simulator, thereby bringing this potential to floating-point devices. This new special-purpose microprocessor is one of the most powerful DSP devices in existence. And perhaps more importantly, the software simulator for the DSP96002 provides the "hooks" so that it is now possible (for the first time) for multiple-processor floating-point systems to be completely and accurately simulated with single clock-cycle granularity, where each individual pin of the device can be fully supported.

1.2 Method of Attack

This text addresses two important problems. The first is the need for DSP systems with higher levels of both performance and fault tolerance. The second is the need for methods to design and develop new multiprocessor and distributed systems in a way that reduces the escalating costs of this extremely complex task.

In addressing these two problems, three objectives are identified. The first objective is the creation of a new system for DSP which provides higher levels of performance and fault tolerance over traditional single-processor or non-redundant DSP designs. This objective is achieved by first considering the requirements of DSP applications, and then correlating them with fault tolerance and performance issues to produce the design for a new system, in terms of both hardware (architecture) and software (operating system communication primitives).

The second objective is the implementation of a new system prototype without the prohibitive costs normally associated with such designs. This objective is achieved by employing state-of-the-art simulation techniques to build a system prototype completely in software.

Finally, the third objective is to evaluate this new system with respect to both performance and fault tolerance. This final objective is achieved by using the system prototype developed in software, along with DSP applications such as digital filtering and fast Fourier transformations, to compare the new system with a uniprocessor system (for performance comparisons) and with a comparable non-redundant system (for reliability comparisons) using the same basic microprocessor.

Based on these objectives, a number of major issues have been considered in the design and development of the system to be described. The second chapter is a background overview of fault-tolerant computing, including such issues as basic definitions, redundancy techniques, communication architectures, properties, and examples of previously designed fault-tolerant systems. The third chapter is an overview of parallel computing, including such issues as parallel processing overview and definitions, parallel architecture models, operating system overview, and real-time computing overview.

In the fourth chapter, an overview of digital signal processing, basic algorithms, DSP microprocessors, and the Motorola DSP96002 device is presented. These issues, along with those in two previous chapters, are then used to discuss and develop the implementation requirements for basic DSP operations, as well as the most effective architecture to address these requirements.

In the fifth chapter, a set of premises and goals are identified for the system, and these parameters are used as the basis for the design and development of a new fault-tolerant computing system. Topics include such considerations as DSP implementations, redundancy techniques, communication architectures, clock synchronization, communication primitives, interface and communication design, and system design enhancements.

The sixth chapter describes how highly accurate simulation is used to essentially develop a prototype for this system entirely in software. This simulation provides an exceptionally accurate description of system behavior. The simulation is used as the basis for the seventh chapter, which describes and quantifies a number of system tests and evaluations. This chapter also includes a discussion on reliability modeling which provides a more analytical description of the fault-tolerant capabilities of this system.

The eighth and final chapter provides the conclusions drawn from this research. Issues addressed include such considerations as system performance, fault tolerance, and growth potential. An overall research summary is included, with consideration of such key points as: problems or questions addressed by this research; objectives; approaches used to achieve these objectives; work performed; and important results and conclusions.

Finally, a set of appendices is included with listings of the software developed for this text. This includes C code for multiprocessor simulation and DSP96002 assembly-language code executed on the processors themselves.

Chapter 2
FAULT-TOLERANT COMPUTING

This chapter provides a background overview of fault-tolerant computing, including such issues as basic definitions, redundancy techniques, communication architectures, properties, and examples of previously designed fault-tolerant systems.

2.1 Basic Definitions

As the research area of fault tolerance has developed, a number of basic terms have been introduced. Of these, the following are some of the most fundamental definitions that will be referenced, as presented by [JOHN89a]:

Availability — is the probability that, at any given instant in time, the system is functioning normally and available for use.

Dependability — describes the overall quality of service provided by the system.

Error — is the occurrence of an incorrect value in some unit of information with the system.

Failure — is a deviation from the expected performance of the system.

Fault — is a physical defect, flaw, or imperfection occurring in the hardware or software of the system.

Fault Avoidance — is the process whereby attempts are made to prevent hardware failures

	and software errors before they occur in the system.
Fault Containment	is the process of confining or restricting the effects of a fault to a limited scope or locality of the system (similarly for errors).
Fault Masking	is the process of preventing faults from causing errors.
Fault-Tolerant Computing	is the process of performing computations, such as calculations, on a computer in fault-tolerant manner.
Fault-Tolerant System	is one that is capable of correct operation despite the presence of certain hardware failures and software errors.
Maintainability	is the probability that, within a specified time, an inoperable system will be restored to an operational state.
Performability	is the probability that, for a given instant in time, the system is performing at or above a specified level of performance.
Reliability	is the probability that, throughout an interval of time, the system will continue to perform correctly.
Safety	is the probability that the system will either perform correctly or discontinue in a well-defined and safe manner.
Testability	is the ability to test for certain attributes or conditions with the system.

2.2 Fault Tolerance Overview

A fault-tolerant system is one that is capable of correct operation despite the presence of certain hardware failures and software errors [JOHN89a]. The need for computers capable of operating despite failures of its components has existed since vacuum tubes and relays formed the hardware component of computers. Historically, fault-tolerant systems have been exemplified by highly expensive systems designed for use in high-profile applications such as spacecraft, aircraft, and other devices where failure of the system could lead to failure of a critical mission and even loss of human life.

This is not to say that features such as high reliability are only required or desirable for such critical systems. Clearly, every user of every computer-based system desires a system which does not fail. But, since the basic premise behind all fault-tolerant systems is redundancy, the cost of providing this capability has been heretofore prohibitively expensive. However, as technology has continued to evolve and develop, this expense has gradually decreased. As the costs of computing technology decrease, the potential applications of fault-tolerant computing increase.

It is now becoming possible for fault-tolerant techniques to be applied to virtually all computing applications. Additional hardware and software can be applied to overcome the effects of a malfunction. This redundancy can range from replicated hardware, with a voting mechanism determining proper results via a quorum, to additional software to repeat the execution of a task in a different way when an error occurs. However, it is impractical to provide complete redundancy for every conceivable contingency, and thus decisions must be made to determine the most likely failures and mechanisms to contain them. One common technique is to design for tolerance of a single failure, which is based on the premise that the probability of multiple failures within a given period of time is extremely small. This time interval can be in terms of how long it takes to repair a failed component or simply how long it takes to recover from a single failure. In addition, many failures are not permanent but are instead transient or intermittent, and thus a simple retry of the previous operation may alleviate the problem [SING90].

The challenge to design and build systems capable of operating in an environment of potential failures is more complex than might be anticipated. It has been historically difficult to statistically describe the type and frequency of failure for hardware faults and software errors or bugs. While models have been developed to describe permanent hardware failures, it becomes even more difficult to forecast the possibility of transient and intermittent failures.

Once a set of fault-tolerance metrics has been decided upon, the choice and possible permutations of techniques for providing tolerance of failures is extensive. Everything from coding theory techniques for correcting errors in memory to extremely complex distributed and massively parallel computing systems may be considered. And, by combining techniques to meet different application requirements, the possibilities are almost endless.

Meanwhile, as technology has progressed, it has become even more difficult to even be sure when a component is fault-free. For example, microprocessors consisting of over one-million transistors are now readily available which are virtually impossible to test completely. At the same time, software is also becoming more complex, making it increasingly more difficult to provide verification and validation, if not impossible. And, to make matters even worse, we need to consider as many potential failures as possible, since a simple oversight might render an otherwise reliable system failure-prone.

While the problems are complex, the rewards are even more desirable. Fault-tolerant design, coupled with continuing advancements in hardware and software technology, has the potential to revolutionize the design of all future computing systems. In fact, future high-end *and* low-end computing systems may one day routinely operate for periods of years without any system failure or downtime.

2.3 Redundancy Techniques

One of the most important principles in any fault-tolerant system is the incorporation of redundancy. Adding redundancy to a system simply consists of augmenting that system with additional resources, information, or time beyond the normal amount required for operation. The different types of redundancy used in fault-tolerant systems can be divided into four categories. These categories are hardware, software, information, and time redundancy.

2.3.1 Hardware redundancy

Hardware redundancy is characterized by the addition of extra hardware resources, such as duplicated processors, generally for the purpose of detecting or tolerating faults. Systems employing hardware redundancy generally use one of three approaches: static, dynamic, or hybrid hardware redundancy [JOHN89a].

Static hardware redundancy techniques use fault masking to hide faults and prevent errors from occurring due to faults. In this way, static designs inherently tolerate faults without the need for reconfiguration or explicit fault detection and recovery. The most widely used form of static hardware redundancy is triple modular redundancy (TMR). With TMR, the hardware is triplicated and the output is determined by taking a vote and choosing the majority or quorum value. The voting mechanism itself can be in the form of a hardware device or software polling routine, and can occur at multiple points in the system. The generalized form of TMR is referred to as N-modular redundancy (NMR), where N is usually an odd integer greater than or equal to three.

Dynamic hardware redundancy techniques attempt to detect and remove faulty hardware from operation. In this way, dynamic designs do not mask faults but instead use fault detection, location, and recovery techniques to reconfigure the system for further operation. The dynamic approach is usually used in systems capable of tolerating temporary erroneous results while the system experiences reconfiguration. Four types of dynamic hardware redundancy are duplication with comparison, standby sparing, pair and a spare, and watchdog timers.

With the duplication with comparison scheme, the hardware is duplicated in order to perform the same operations in parallel. The results of these parallel operations are compared, and if they differ, an error condition is generated, thereby providing a level of fault detection. In the standby sparing scheme, one module operates while one or more modules serve as standby spares. When a faulty module is detected and located, it is replaced in operation by one of the spare modules during system reconfiguration.

The pair and a spare scheme is essentially a combination of the duplication with comparison and standby sparing schemes. That is, two modules operate in parallel, each of which having one or more standby spares available for replacement in the event of a module fault. In the event of a fault, one or both of the operating modules are replaced, depending on the strategy. Finally, the watchdog timer scheme is sometimes an effective technique for detecting faults, and consists of timers which must be accessed within a prescribed amount of time. In this way, any hardware component not responding for an excessive amount of time can be considered faulty. Watchdog timers can also be used to detect software errors, such as an infinite loop which would prevent prompt timer access.

By combining some of the schemes discussed, it is possible to incorporate features of both the static and dynamic hardware redundancy techniques. This is known as hybrid hardware redundancy. Some of these combinations are NMR with spares and triple-duplex architecture, the latter of which being a combination of TMR and duplication with comparison. These techniques tend to require significantly more hardware resources, and are usually applied to systems that require an exceptionally high degree of computational integrity.

2.3.2 Software redundancy

Software redundancy is characterized by the addition of extra software resources beyond that required for normal system operation, generally for the purpose of detecting or tolerating hardware faults or software errors. Software redundancy can involve completely replicated programs, or just the addition of a few lines of code. The three major software redundancy techniques are consistency checks, capability checks, and software replication [JOHN89a].

Software consistency checks consist of verification methods used to check the characteristics of information generated and used by the system. This may include checking the range of resulting values, comparing the performance of the system with some predicted performance, or overflow checks. Software capability checks are used to verify the capability of a component of the system. This may include memory tests, processor tests, or communication tests. Both consistency and capability checks are used to detect faults that may occur in the hardware of the system.

With software replication, the goal is to detect and potentially tolerate errors that may occur in the software for the system. Unlike hardware, software does not fail in the normal sense of a component failure. Instead, software errors are the result of incorrect software requirements and specifications or programming mistakes (i.e. software design mistakes). Thus, a simple duplication with comparison technique would be useless in detecting a software error, since each of the duplicated modules would run the same software and thus suffer the same software errors.

Two approaches to software fault tolerance via software replications have emerged. They are N-version programming and recovery blocks. With *N-version programming*, the software module is designed and programmed by N separate and independent groups of programmers, often using different algorithms and programming languages. While each group works from the same specifications, the hope is that they will not make the same mistakes. Then, during implementation, all N software modules are executed, and the N results of the N software modules are compared to choose the output value. Of course, care must be taken in the specifications, otherwise all the groups will reproduce faulty software based on specification errors [CHEN78].

While N-version programming provides a static redundancy approach, *recovery blocks* represent a dynamic redundancy approach to software fault tolerance [PRAD86a]. The recovery block structure consists of three software elements. First, a primary software routine is used to execute critical software functions. Second, an acceptance test is used to test the output of the primary routine after each execution. Third, one or more alternate routines are used to perform the same function as the primary routine and are invoked by the acceptance test when an error is detected in the primary routine. Unlike N-version programming, only a single implementation is run at a time, and the result is decided by a test instead of a comparison of functionally equivalent alternate routines. The alternate routines are deliberately chosen to be as uncorrelated to the primary routine as possible. Of course, one of the most critical concerns of recovery blocks is the design of the acceptance tests, since determining what is "acceptable" can be extremely difficult [RAND75].

While software fault tolerance via software replication has received much attention in recent years, it is important to note that software redundancy is still in its infancy relative to other forms of redundancy. Software has become by far the most expensive component in the life cycle of a modern computing system, and thus these techniques are extremely limited in their applicability. That is, while hardware resources have continued to decrease in cost as technology has evolved, software costs have continued to increase. Historically, only the most critical systems requiring maximum reliability (despite the high cost) have warranted use of software replication methods like N-version programming and recovery blocks. Though this situation may change, at least at this point in time software replication is extremely limited in its cost-effectiveness compared to other redundancy methodologies.

2.3.3 Information redundancy

Information redundancy is characterized by the addition of extra data or information to provide a capability for fault detection, masking, or tolerance. The most notable approach to information redundancy stems from the reliable communication arena in the form of coding theory. For example, by adding a certain number of redundant bits to each data word, it is possible to develop codes that can detect and even correct a fixed number of bit-level errors. Information redundancy can be used in a variety of ways in a computing system, such as error-detecting and error-correcting memory subsystems.

Due in part to its origins and extensive study in communication systems, descriptions of the mathematical foundations of coding theory for computer systems are widespread. An excellent introduction is provided in [PRAD86a] and [RAO89].

Another approach to providing fault detection and fault tolerance that incorporates elements of both hardware redundancy and information redundancy is the algorithm-based fault tolerance technique [ABRA87]. This technique is associated with VLSI circuits made up of many extremely simple processors which are dedicated to one particular application or algorithm. The most widely studied algorithm for which this fault-tolerance technique has been applied is the problem of matrix multiplication.

With matrix multiplication, each of the independent processors is dedicated to computing one of the elements in the resulting product matrix. Information redundancy is applied in the form of a matrix checksum row and a checksum column in each of the input matrices and the product matrix. Hardware redundancy is inherent, since extra processors are used to compute the extra checksum elements in the product matrix. In the event that one of the elements in the product matrix is erroneous (e.g. perhaps due to a failure in the element's independent processor), the checksum row and column are used to detect the location of the element and to determine its correct value.

2.3.4 Time redundancy

When considering a mechanism for redundancy, the resources required are always of paramount importance. While both hardware and information redundancy techniques can require a large amount of extra hardware for their implementation, sometimes this investment is not necessary. In order to reduce the need for extra hardware, the time redundancy approach has emerged. By using an additional amount of time to perform the given task, it is possible to provide similar capabilities as some hardware and information redundancy implementations. This can be a useful alternative, since many applications are less dependent upon time than they are hardware related issues, such as cost, weight, size, and power consumption [JOHN89a]. For example, in the control system for a guided missile, the processor or processors may have a large amount of reserve computational power but a critical need for minimal weight and size.

The main idea of time redundancy is to perform computations a number of times in order to provide fault detection. Time redundancy may be used alone or in concert with other redundancy techniques. For example, information redundancy (via error-detecting codes) could be used for error detection, and time redundancy could then be called upon to distinguish between permanent and transient errors via repetition.

The main difficulty with the time redundancy approach is making sure that the system has the same data to operate on during each repeated computation. For example, if a transient fault were to occur, it might corrupt the data that a subsequent repeated computation would be based on. However, there appears to be great potential for the detection of permanent faults via time redundancy [JOHN89a]. Of course, this is only the case for those applications that can afford the extra time involved.

2.4 Fault-Tolerant Communication Architectures

One of the most important considerations in a fault-tolerant computer is the choice of an architecture for communication between nodes (i.e. processors or memory). Many issues must be considered in making this selection, such as communication bandwidth, latency, redundancy, and isolation. As a starting point, we will consider Pradhan's seven categories of fault-tolerant multiprocessor communication architectures as described in [PRAD86a]. These are:

- reliable shared buses
- shared-memory interconnection networks
- loop architectures
- tree networks
- dynamically reconfigurable networks
- binary cube interconnection networks
- graph networks

While other architectures are conceivable, they can be described in terms of one or more of the categories listed.

2.4.1 Reliable shared buses

When considering simplicity and cost effectiveness, the shared bus approach to fault-tolerant multiprocessors is very attractive. With this technique, a single bus is shared by all the nodes via time-division multiplexing. However, this clearly introduces a single point of failure to the system, and thus numerous measures must be employed to overcome this vulnerability [PRAD86a].

Whenever access of a single resource is provided to multiple entities, some sort of arbitration is required. This is no different with a shared bus. Only one processor can control the bus at each instant in time, and that processor is known as the *bus master*. This arbitration can be controlled either by a central source (the bus controller) or by distributed sources (the processors themselves).

The primary causes of shared bus failure are threefold: *control* failures; *synchronization* failures; and *bus connection* failures. Control failures consist of problems with bus allocation and deallocation. For example, in the case of centralized arbitration, the failure of the bus controller is catastrophic. Even with distributed arbitration, it is possible that the current bus master might suffer a fault and never relinquish the bus.

Synchronization failures consist of problems with handshaking. For example, a misalignment between clock and data cycles at the receiver is known as *clock-skewing*, and can result in errors [STON82]. Even in asynchronous communications where a clock is not used as part of the transmission, timing problems with handshaking signals between the transmitter and receiver can be troublesome.

Finally, bus connection failures are concerned with faults in the actual bus interface and the bus communication lines themselves. While the bus lines are passive, and thus less likely to fail, both types of faults must be considered. The failure of a bus line would render the bus inoperable, as could the failure of a bus connection between a node and the bus.

Of course, since redundancy is perhaps the key element of any fault-tolerant device, the consideration of redundant buses is natural as a solution to bus connection failures. That is, to allow for failures like those described, one or more extra buses are provided to, in effect, remove the single point of failure implied by the single shared bus. Many permutations of multiple bus systems are possible, in terms of both how they provide interconnection and how they handle failures. In fact, these techniques have evolved into shared-memory interconnection networks, which will be discussed next.

2.4.2 Shared-memory interconnection networks

A common technique that is used in multiprocessors is to share part or all of available memory between all of the processors. These multiple memory modules form part or all of the address space of each processor in the system.

In order for this shared memory to be accessed, four basic types of interconnection networks have arisen for connecting processors and memory modules. These are the *common bus*, *crossbar switch*, *multibus/multiport*, and *multistage* shared-memory interconnection networks [PRAD86a].

Common bus shared-memory interconnection networks consist of a single bus shared among all of the processors and memory modules. While this is the simplest approach, as was seen in the last section, it can also be problematic from a fault tolerance standpoint.

Crossbar switch shared-memory interconnection networks use a single crossbar switch for communication among all of the processors and memory modules. The switch allows simultaneous connection between all non-overlapping processor-memory pairs. The switch itself is responsible for all arbitration and connections, thereby simplifying the access methods used by the processors. However, as the switch complexity grows exponentially with increases in the number of processors and memory modules, it becomes a major single point of failure.

Multibus/multiport shared-memory interconnection networks use dedicated buses or links to individually and separately connect each processor to each memory module. Each of the memory modules is multiported (i.e. each has multiple interface ports). Since typically only one access to each module is possible, arbitration and prioritization logic is incorporated in the multiporting logic of each memory module. Thus, while removing the single point of failure exhibited by the crossbar switch, the memory modules themselves become much more complex.

Finally, multistage shared-memory interconnection networks use a large number of relatively simple switches in a series of stages, each of which consisting of multiple switches. Processors and memory modules are connected to the inputs and outputs of these networks, allowing many possible processor-processor and processor-memory connection pairs (e.g. see Figure 2.1). Some of the choices that determine the construction of these networks are the number of stages, number of switches per stage, the size of the switches, the capabilities of each switch, and how the stages are interconnected.

For example, some networks are referred to as *strictly non-blocking*, whereby any connection can be made between any unused inputs and outputs no matter what existing connections have been made. If a uniform routing routine must have been followed to achieve this capability, these networks are called *wide-sense non-blocking*. On the other hand, some networks are referred to as *rearrangeable*, whereby any connection can be made between any unused inputs and outputs after a temporary delay for network connection rearrangement. And still others are referred to as *blocking*, whereby some connections between unused inputs and outputs cannot be made, depending on existing connections [BROO83].

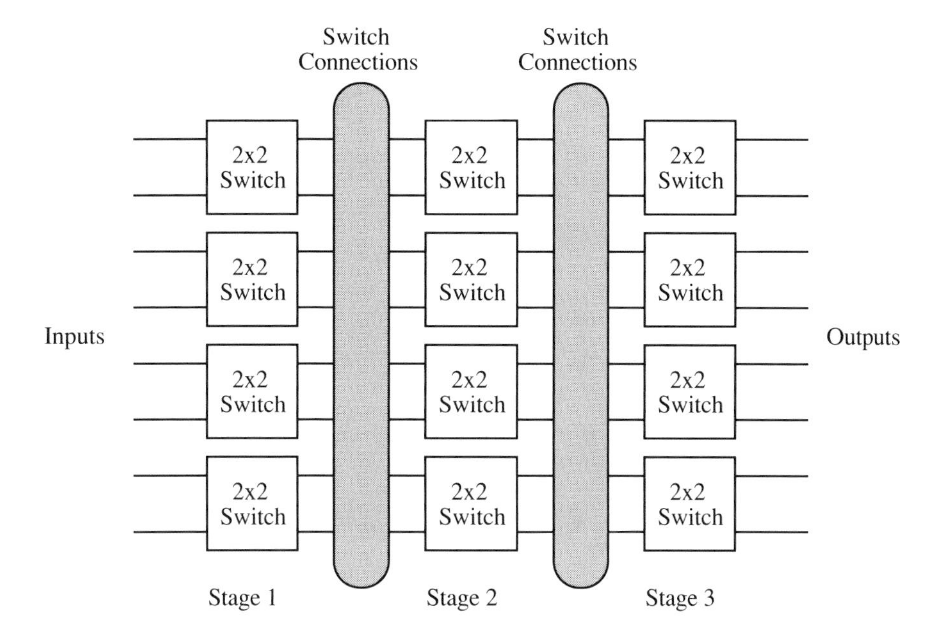

FIGURE 2.1 A Generic 3-stage 8-input/output Network

2.4.3 Loop architectures

Another approach to fault-tolerant communication architectures is perhaps best recognized from local area networking. The loop or ring architecture consists of processors which communicate via a circular loop in a clockwise manner, a counter-clockwise manner, or both. One of the main advantages of these architectures are their ability to withstand a faulty processor or link between processors. For example, if five processors are connected in a loop, a failure in any single processor can be managed by routing traffic between the other processors using the remaining path (see Figure 2.2) [PRAD86a].

Some of the choices that differentiate between loop architectures are the number of processors, the possible directions of data flow, how data is inserted and removed from the loop, how tokens are used to determine loop control, and the provision of extra links between nodes. In addition, loops can be constructed using multiple levels, with loops of sub-loops.

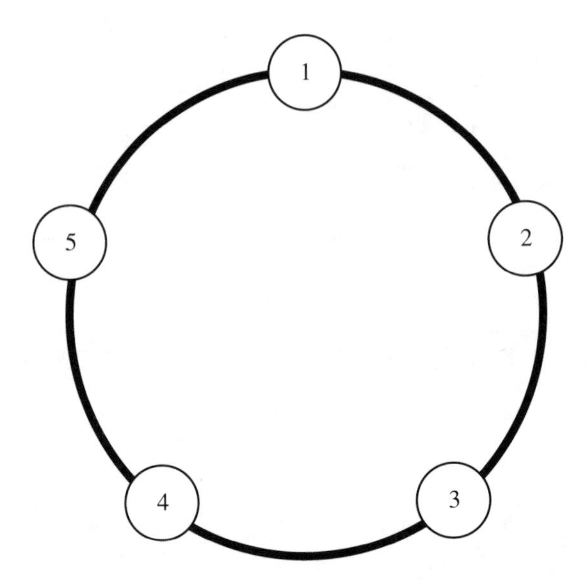

FIGURE 2.2 A Generic 5-processor Loop Architecture

2.4.4 Tree networks

So far, all of the communication architectures that have been considered fall into the category of *horizontal* system organizations. In horizontal systems, each processor exercises independent control, and thus does not fall under the control of any of the other processors. On the other hand, with *hierarchical* system organizations, processors are divided into different levels. In these hierarchical or tree networks, processors at a lower level in the tree are controlled by those at a higher level [PRAD86a].

Many parallel algorithms lend themselves to a "divide and conquer" or hierarchical approach more than a "linear" or horizontal one. Thus, for some applications, tree networked systems have the potential to provide superior performance. Of course, like a real tree in nature, there can be problems with a single point of failure, such as the root.

In order to make tree networks more tolerant of processor and link failures, two basic approaches have been studied. One technique that has been considered is that of augmenting a simple binary tree with extra links so that the tree remains fully connected despite one or more processor or link failures. In this way, although the failed processor(s) may not be available, the rest of the tree is able to continue functioning, thereby providing a graceful degradation in performance [DESP78].

Another technique that has been proposed augments a simple binary tree with extra links and processor nodes. Unlike the first approach, the idea here is to preserve the original tree structure fully, by reconfiguring the tree when one or more failures occur [HAYE76, KWAN81].

2.4.5 Dynamically reconfigurable networks

As previously mentioned, some applications are better suited to horizontal configurations while others to hierarchical ones. In order to achieve the best of both worlds, dynamically reconfigurable networks were devised. These networks are characterized by the ability to change from a horizontal to a hierarchical configuration or vice-versa while the system is operational. In effect, multiple distinct architectures are realized by using the same physical interconnections [PRAD86a].

Pradhan describes a dynamically reconfigurable fault-tolerant (DFT) network architecture for systems with homogeneous processors [PRAD85b]. For a network with n processors or nodes, each node i is connected to both node $(i + 1) \bmod n$ and $(2 * i) \bmod n$. (see Figure 2.3 where $n = 7$).

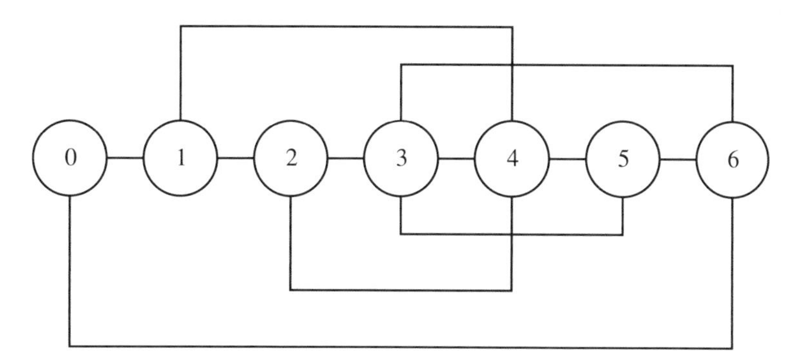

FIGURE 2.3 A 7-node DFT Network

By only making use of the i to $(i + 1)$ connections, this network takes on the form of a logical loop for horizontal applications. For hierarchical applications, the connections for a logical binary tree configuration are slightly more involved. In a binary tree, we can generalize the node numbers so that each node j has left child $(2 * j)$ and right child $(2 * j + 1)$. Thus, in order to provide binary tree connections, each node communicates with its left child using the $(2 * i)$ connection of the DFT and with its right child using a two-step process from the node to the right child via the left child (by using the $(i + 1)$ DFT connection). For example, in the $n = 7$ DFT, node 2 has children numbered 4 and 5. Node 2 can communicate with its left child (node 4) directly, and with its right child (node 5) via the left child (see Figure 2.4 where dashed lines indicate "via the left child").

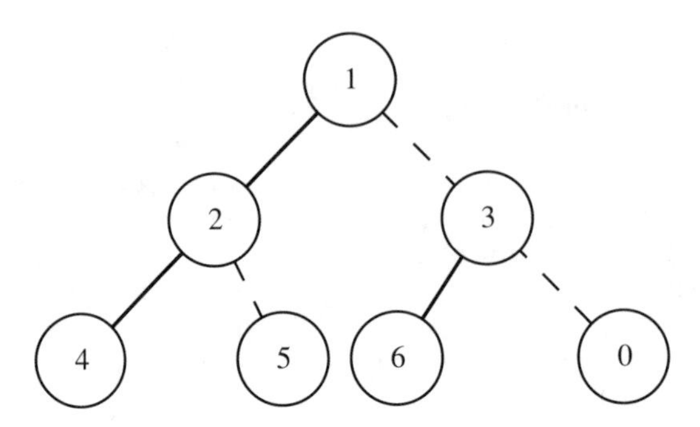

FIGURE 2.4 A 7-node DFT Logically Configured as a Binary Tree

The DFT architecture has inherent fault-tolerance features. In logical loop configuration, the failure of a node or link can be handled, since bidirectional rings are capable of withstanding a single failure (as was discussed in the section on loop architectures). In logical binary tree configuration, algorithms have been developed to reconfigure the logical tree so that the faulty node appears at or near a leaf position, thereby minimizing its effect on tree communication [PRAD85b].

One of the inherent problems with communication in a binary tree configuration is the bottleneck that the root node often becomes. However, using the DFT approach, it is possible to provide more uniform message passing by using all of the connections of each node, and not just the logical tree connections. These techniques have also been extended to better handle faulty nodes or links [PRAD85b].

2.4.6 Binary cube interconnection networks

One communication architecture that has received a lot of attention in recent years is the binary cube or hypercube. In these networks, the number of processors or nodes is always a power of two, and each node has associated with it a binary address of $m = \log_2 n$ bits, where n denotes the number of nodes in the system. With binary cubes, there are exactly m disjoint paths between any two nodes. Thus, it can be shown that this architecture is inherently tolerant of up to and including (m - 1) faulty nodes or links [KUHL80, ARMS81].

In a binary cube (or n-cube), two nodes are adjacent (i.e. exactly one hop apart) if and only if their binary addresses differ by exactly one bit. Thus, the number of bits differing between any two node addresses determines the minimum number of hops in the communication path between them, thereby providing a relatively simple routing algorithm. For example, consider a network with $n = 8$. Since $m = 3$, this means that any pair of nodes can communicate with each other at a cost of at most three hops, and the failure of up to two nodes or links can be tolerated while still providing communication between all possible functioning nodes (see Figure 2.5).

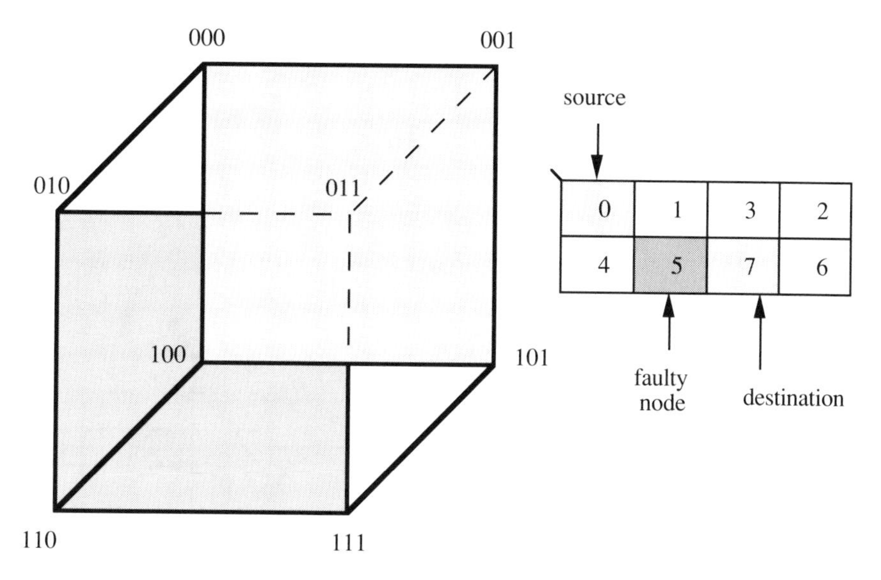

FIGURE 2.5 An 8-node Binary Cube and the Presence of a Faulty Node

For a message to be passed from node 0 to node 7, there are three possible disjoint paths. They are 0-4-5-7, 0-1-3-7, and 0-2-6-7 (i.e. 000-100-101-111, 000-001-011-111, and 000-010-110-111 in binary respectively). If node 5 is faulty, either of the latter two paths are still available for communication.

Another interesting aspect of this network is that the number of nodes can be doubled simply by adding an extra connection per node. However, for larger systems, the extra connections required tend to be a limiting factor. A closely related topic of research that has been studied is the development of traversal and diagnostic algorithms for the binary cube in the presence of faulty nodes and links [HAWK85].

2.4.7 Graph networks

This final category of fault-tolerant communication architectures is based on certain types of graphs. These fault-tolerant graph or FG networks combine some of the more attractive features from some of the networks described earlier. Using various nondirected graphs as models, alternate multiprocessor multibus configurations have been proposed [PRAD83, PRAD85a, PRAD86b].

A nondirected graph is a pair of sets. The first is a set of vertices or nodes. The second is a set of edges or arcs which describe connections between nodes, where no implied direction or ordering is implied by the edges. Graphs provide a geometric or pictorial aspect to the study of communication architectures [MOTT86].

In general, two different approaches have been proposed: a link-based configuration and a bus-based configuration. Consider a typical graph with some number of nodes and edges. With the link-based approach, the nodes are treated as processors and the edges as links between processors. Alternately, with the bus-based approach, the nodes are treated as bus connections and the edges as processors (see Figure 2.6).

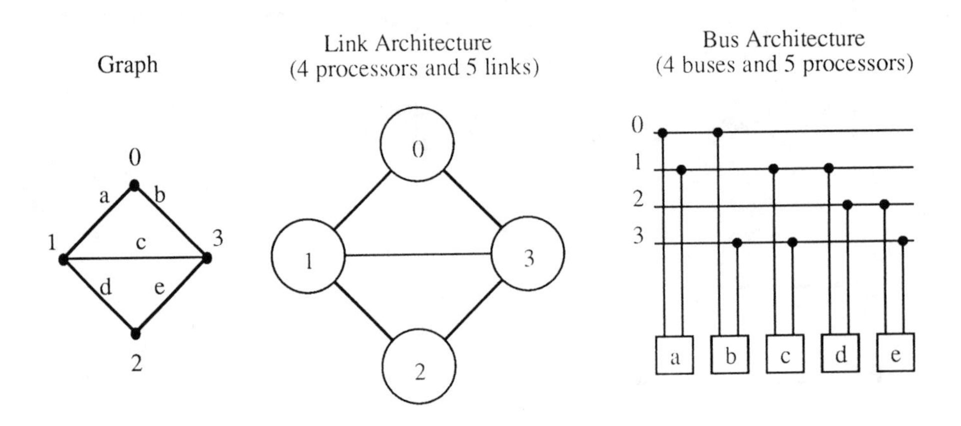

FIGURE 2.6 An Example Link and Bus Architecture

One interesting aspect of these two approaches is the correlation between them in the presence of faults. A faulty processor in the link architecture is comparable to a faulty bus in the bus architecture. And, a faulty link in the link architecture is comparable to a faulty processor in the bus architecture. The main variable in these types of networks is the type of graph chosen as the model for the system and the fault handling ability of each approach based on that model.

2.5 Atomic Transactions

Any system intended to provide fault tolerance which incorporates disk file I/O must address the problem of file system update in the presence of faults. The model commonly used is one taken from magnetic tape transaction processing. Using this model, a master tape is read in, along with update tapes, in order to produce a new master tape as output. If the system fails during the update, the original master is used to begin the update again. If the update completes without a failure, the new output tape becomes the master tape, and the process continues again at a later time.

This approach is referred to as *atomic transactions*. Essentially, in order to update an element, the transaction is either completed in its entirety or not at all. No intermediate modifications are to be allowed. One technique that may be applied to provide this mechanism is described by Lampson, and consists of multiple levels of abstraction [LAMP81].

The first and lowest level of abstraction is referred to as *careful disk*. Every write operation consists of a CAREFUL_WRITE, which performs the write followed by an immediate read verify. If the CAREFUL_WRITE fails beyond a certain number of tries, the block is marked *bad* and an error is returned signifying failure. By using this write with verify technique, a CAREFUL_READ operation is almost guaranteed to be successful, unless of course the block has gone bad since it was written, which is considered highly improbable.

The next higher level of abstraction is referred to as *stable storage*. A stable storage block consists of a pair of associated careful blocks typically on different disk drives. The STABLE_WRITE operation consists of a CAREFUL_WRITE on the first of the careful blocks in the pair, followed by the other. If a failure occurs during the first CAREFUL_WRITE, an error is returned, the first block is marked *bad*, and the second careful block is left unmodified. If however a failure occurs during the second CAREFUL_WRITE, an error is returned, the second block is marked *bad*, and the first careful block is left correctly updated.

At periodic intervals, a support program executes and evaluates all stable storage blocks. If both careful blocks are marked *good* and are identical, then the stable storage block is left unmodified. If one careful block is marked *good* and the other *bad*, the *bad* one is updated from the *good* block. And, if both careful blocks are marked *good* but contain different data, the second block must be stale and therefore is updated from the first block. In this way, disk blocks are atomically updated and tolerant of system faults [TANE85].

2.6 Communication Network Properties

A set of properties has been described by Garcia-Molina and Pittelli to model both the communication network and the processors of a distributed computing system [GARC89]. These network properties are: no transmission errors (if the message is received, then it is received correctly); no spontaneous messages (if the message is received, then it was sent previously); messages delivered in the order sent (message ordering is preserved); and verifiable origination (the receiver of the message is able to verify the source of origin).

If a network exhibits these four properties, then it is referred to as *trustworthy*. In addition, if it has the property of secure messages (a message cannot be altered without detection), then the network is said to be *secure*. And, if the network has guaranteed delivery (a message sent will be received if the receiver is active), then it is said to be *dependable*.

Similarly, a set of computing node or processor properties can also be considered. The node properties are: safe storage cells (memory is atomically updated); fail-stop (processor failure implies immediate halting of all processes); and identical recovery algorithms (all nodes use the same approach).

If a node exhibits these three properties, then it is referred to as a *sane* node. In addition, if the node has the property of timely execution (the node always responds to incoming messages within a certain time period), then the node is said to be *dependable*. And, if the node has no failures (the node always performs as designed), then it is considered *infallible*.

In order to formulate the algorithms of a fault-tolerant distributed system, these properties must be considered. For example, if a node can become *insane*, then it could potentially propagate invalid information throughout the entire distributed system. Recovery techniques for systems with potentially insane nodes must provide mechanisms to prevent this information contamination. A five step process for dealing with the failure of insane nodes is: replicate the processes to be accomplished; transmit the input data to the duplicate nodes; execute the process on each duplicate node independently and in parallel; collect the output from the duplicate nodes; and compare the output values and consider the majority result as the correct output.

2.7 Examples of Fault-Tolerant Systems

A number of fault-tolerant computing systems have been developed by researchers. Each of these systems can be classified in terms of their fault-tolerance goals. These goals are high-availability, long-life, and critical-computation. Although not specifically designed with DSP applications in mind, we will overview some systems that will help provide a frame of reference for developing a new system.

2.7.1 Tandem Non-Stop computer system

The Non-Stop line of fault-tolerant computing systems was introduced by Tandem in 1977. The Non-Stop systems were designed for high-availability commercial on-line transaction processing systems, including such applications as banking transactions and airline reservations. While representing the first successful commercially available fault-tolerant computing system, these systems were originally designed to provide at least an order of magnitude improvement in availability compared to existing commercial systems, using off-the-shelf components [DIMM85].

Due to the high expense of processors at that time, dynamic redundancy was chosen over static redundancy. The key design objectives were: no single hardware failure should stop the system; hardware elements should be maintainable on-line; database integrity should be insured; and the system should be capable of undergoing hardware additions without software changes. The basic premise behind the design is one of complete duplication. This includes: two independent loosely-coupled minicomputers; two independent data paths to everything; dual-ported peripheral controllers; and two separate power supplies.

A dual interprocessor bus is employed, called the *Dynabus*, which features electrical isolation and a message passing orientation. In effect, the Dynabus forms a local area network (LAN) of between two and sixteen processors per system. Each processor includes a power supply, Dynabus interface logic, a microprogrammed CPU supporting virtual memory and floating-point arithmetic, main memory, and an I/O channel. Dual-ported I/O device controllers are used, where each is attached to any pair of processors. Similarly, dual-ported disk drives are used, where each is attached to any pair of I/O controllers.

Control circuitry is provided to disconnect each processor from an errant Dynabus, and I/O controller circuitry is provided to disconnect each I/O controller from an errant I/O bus. Similarly, disk drive control circuitry is provided to disconnect each drive from an errant I/O controller, and disks can be pairs in shadow sets. Individual processors, I/O controllers, and power supplies may be powered down during operation to provide graceful degradation while repairs are being effected.

In terms of software features, the system is basically a message-passing LAN-based multicomputer, as opposed to a multiprocessor or distributed system, where each processor runs a separate copy of the operating system. This makes the operating system simpler to implement and can provide more fault isolation. The requirements for the operating system were that it should be: operational despite a single failed module or bus; operational during the repair of any failed module or bus; implemented in a reliable manner; support all conceivable hardware configurations; and keep the physical configuration hidden from user applications.

All processes are executed as identical pairs, where each runs on a different processor (one primary and one backup process). The primary process actively executes, and periodically sends checkpoint information to its quiescent backup process. Each processor broadcasts an "I'm alive" message over the Dynabus to the other processors every second, and if after two seconds there is no response from a processor, the others consider that processor as failed, and they dispatch their appropriate backup processes as new primary processes. Later, when the failed processor returns to service, new backup processes are then created on it [DIMM85].

2.7.2 Stratus computer system

The Stratus S/32 Continuous Processing System, developed by Stratus Computer Incorporated, is a fault-tolerant system designed to compete with Tandem in the high-availability on-line transaction processing arena. This system was first delivered in 1982, is also marketed by IBM as the System/88, and was designed with an emphasis on reliable data storage coupled with continuous availability [LEE90].

The primary method by the Stratus system for fault-tolerant computing is based on static hardware redundancy. Based on the Motorola 68000 microprocessor, up to thirty-two processing modules are supported. Each module consists of either one 68000, memory controller, disk controller, communications controller, and a 1.4 MB/sec intermodule link (in simplex configuration), or two of each (in duplex configuration), as well as a tape controller [WILS85]. In addition, every processing module board also runs in duplex mode with another to form a board-level static redundancy. In this way, each process runs simultaneously on four processors [LEE90].

In contrast to the Tandem system where a large part of the fault tolerance is implemented via software redundancy and dynamic hardware redundancy, the Stratus system uses static hardware redundancy coupled with self-checking circuitry for component-level error detection. The effect of this difference is that a much larger part of the system resources is dedicated by the Stratus system in providing fault tolerance, since as many as four processors are performing identical tasks. However, the simplicity of this approach becomes more attractive as the hardware resources continue to decrease in cost.

2.7.3 Electronic switching system

The Electronic Switching System (ESS) is a series of systems developed for the high-availability application of telephone call switching. Since the first version was introduced in 1965, the basic objective of each system has been to minimize both down-time and incorrect handling of phone calls, as well as to detect and locate faults as quickly as possible in order to assist maintenance and repairs [JOHN89a].

The basic redundancy technique of the ESS systems is duplication using dynamic hardware redundancy. That is, all of the functions of each processor are duplicated, and the detection of a fault in any part of a processor results in that processor being switching out of operation. However, instead of comparing the results of duplicated processors in order to provide fault detection, each processor is designed with self-checking circuitry, thereby allowing the processors to operate without strict synchronization.

Processor redundancy is handled via a standby sparing approach. If and when the operating processor detects its own fault, the spare processor begins executing and assumes the responsibilities of the processor it is replacing. In support of this self-checking capability, information redundancy is employed in the form of error-detecting codes for address and control data in the microprogram store, as well as the main memory and registers.

2.7.4 Space Shuttle computer system

Perhaps one of the most widely recognized applications of fault-tolerant computing is the space shuttle computer system. Designed as a critical-computation system with an emphasis on reliability, the shuttle computer system is responsible for flight-critical functions during ascent, reentry, and descent. It consists of five identical general-purpose computers, each of which capable of performing both critical and noncritical operations. During flight-critical situations, four of the computers are configured in an NMR form (where $N = 4$) with voting, while the fifth performs noncritical functions in a background role [JOHN89a].

The system itself was based on five requirements. First, each computer must use self-test features which allow it to detect 96% of all possible faults within that computer. These tests are primarily implemented in software, since off-the-shelf general-purpose computers were to be used. Second, the failure of two computers in the NMR set must be identified to the crew, and as many as three when possible. Third, each computer must cease all transmission during a failure, in order to prevent erroneous data from being propagated to the other computers.

Fourth, fault isolation must be provided to minimize the extent to which a failure in one computer can result in a failure in another computer. And finally, transient fault recovery should be provided when practical, to minimize the impact of transient faults on the reliability and performance of the system.

Communication between the processors and the I/O devices is provided by twenty-eight serial data buses, with five of these buses being dedicated for interprocessor communication. Fault detection is primarily accomplished using built-in-test and self-test features, bus time-out tests, comparisons, and watchdog timers. Overall, the shuttle system designers have reported that this combination of fault detection techniques has provided for better than the 96% level of fault coverage desired [SKLA76].

2.7.5 Fault-tolerant multiprocessor

Another fault-tolerant system developed for critical-computation applications is the Fault-Tolerant Multiprocessor (FTMP). The FTMP system was designed to provide an extremely reliable platform for aerospace applications such as commercial transport aircraft. The basic requirement of the FTMP system was to achieve probability of failure (i.e. the reciprocal of reliability or unreliability) of 10^{-9} or less during a ten-hour flight. This value was chosen to match that provided by the mechanical components in a typical modern aircraft [JOHN89a].

The designers of the FTMP system felt that the crucial parts of the system could be best achieved in an economical sense by using hardware, and thus the FTMP system is very much a hardware-intensive approach to fault tolerance. Based on a conventional shared-bus multiprocessor, the system consists of processors formed in triads, system memory, a real-time clock, a control unit, input/output units, and the system bus.

The processors in the FTMP system are divided into TMR processor triads with voting. That is, they are divided into groups of three, and each group operates using the exact same software in tight synchronization via a fault-tolerant clock. These triads operate in parallel with each other, and processors not participating in a triad act as spares. When a processor fails in a triad, it is replaced with a spare. However, if a spare is not available, all the processors in the triad are withdrawn from service, with those still functioning serving as spares for the remaining triads.

The system bus consists of five serial buses. Three of the buses form a TMR bus triad with voting, while the other two serve as spares. In turn, even memory modules are configured in a TMR triad, so that communications between processor and memory triads is voted upon at every stage in the transfer.

2.7.6 Software implemented fault tolerance

Designed to be directly competitive with the FTMP system, the software implemented fault tolerance (SIFT) system was also intended for commercial aircraft control with an unreliability of at most 10^{-9} during a ten-hour flight. Prototypes of both the FTMP and SIFT systems were funded by the NASA Langley Research Center and delivered in early 1983 [JOHN89a].

As their names imply, the SIFT and FTMP systems differ in their fundamental approach, with the former emphasizing a software-intensive technique and the latter a hardware-intensive one. The advantages of the software-intensive approach lies both in its ability to support off-the-shelf hardware (as opposed to application-specific hardware) and its flexibility (due to the relative ease of software modification compared to hardware changes). However, since a significant amount of the processing capability of the system must be used to perform fault-tolerance functions, the software-intensive approach can require more processors to achieve a comparable level of performance.

The designers of the SIFT system based their design on four basic premises. First, they wanted to maximize the use of off-the-shelf standard hardware components. By doing so, they felt that reliability could be increased, since they felt that component reliability increased with the maturity of the device and larger production. Second, the system should avoid dependencies associated with shared elements as much as possible, in the hopes of reducing single points of failure. Third, all system software should be written in a high-level language to decrease coding errors and increase modularity. Finally, they considered simplicity of design to be a key component of the overall system, since simpler designs are more easily verified as well as understood when considering potential faults.

Each of these four premises are manifested in the SIFT design. The smallest units of hardware are conventional computers composed of off-the-shelf processors and memory, so that a fault in any part of a computer causes the system to completely remove that computer from operation during reconfiguration. The only shared elements in the SIFT architecture are the computer intercommunications bus and the primary power supply, and both of these are designed to tolerate faults. All system software was written in a high-level language, and the system design was simplified by using majority voting on NMR computers as the sole means of fault detection, containment, and masking.

In the SIFT architecture, all computers are interconnected via point-to-point links called the computer intercommunication bus, and a shared dual-redundant pair of serial I/O buses are used to provide fault-tolerant communication between I/O devices (e.g. aircraft sensors and actuators) and the computers. Since the SIFT system is a software-intensive implementation, some of the most important aspects of the system deal with system software.

The executive software performs a number of functions. It schedules and controls all tasks in the system and performs voting for fault masking. It detects and diagnoses faults in the system, and reconfigures the system to remove faulty hardware components. It ensures that task output results are transmitted to all computers in a timely fashion, and ensures that identical copies of the input data is provided to the computers. Finally, it is responsible for the input and output functions to ensure that all sensor data is received and actuator data is transmitted.

The design of the SIFT system addressed a number of challenging problems. Some of these include distributed clock synchronization, determination and consensus of system health in the presence of faults, and mathematical proof of software correctness [WENS78].

2.7.7 August Systems industrial control computers

Another system designed for critical-computation applications is the August Systems CS-3001 fault-tolerant control computer. Intended for critical process control and monitoring in an industrial environment, the CS-3001 system uses TMR to achieve fault tolerance and high reliability using design concepts like those of the SIFT system [AUGU86].

The system is made up of a control computer module (CCM), a dataport module (DPM), and a peripheral interface module (PIM). The CCM consists of a TMR set of Motorola MC68020 processors connected by internal point-to-point interprocessor communications links. As a TMR set, these processors are tightly synchronized, operate on the same data using the same software, and use voting to determine results. The DPM consists of a TMR set of bus adapters connected to the CCM by a TMR set of buses. Finally, the PIM provides the interface from the bus adapters in the DPM to the I/O devices.

In addition to the TMR format normally used with this system, the architecture provides the flexibility to use less than triple redundancy (e.g. duplex or simplex) when the application requirements warrant it. During operation, input data arrives over the triplicated buses. Each processor receives the information from all three buses, and performs its own majority vote to determine the value it will operate upon. Results of the computations by all of the triplicated processors are provided over the triplicated buses to the DPM, which performs a majority vote and provides this result to the appropriate I/O device interface [JOHN89a].

2.7.8 The C.vmp system

Another system for critical-computation and high-reliability real-time control applications is the computer voted multiprocessor or C.vmp system. With the C.vmp system, the designers hoped to be able to tolerate permanent and transient faults, provide fault survival in a way that is transparent to user software, avoid any lost time due to fault recovery, use standard off-the-shelf hardware components in a modular fashion, and in general be able to dynamically balance performance and reliability tradeoffs [PRAD86a].

The C.vmp system is in effect a multiprocessor system, based on the LSI-11 microprocessor, which is capable of fault-tolerant operation using a TMR approach with hardware voting. The redundant hardware can operate either independently or, by way of an external event or internal request, can be reconfigured into a fault-tolerant mode. For the latter mode, in order to maximize modularity and transparency to user software, the designers chose bus-level hardware voting as the primary fault-tolerance technique. In this way, voting takes place every time the triplicated LSI-11 processors either write to or read from the triplicated bus which connects them with the triplicated memory modules.

When the voter is in operation, disagreements in triplicated processor or memory module outputs can be prevented from propagating throughout the system. The voter is a memoryless hardware component which may or may not be totally triplicated, depending on application requirements. The voting takes place in parallel on a bit-by-bit basis, so that the operation can continue as long as at least two of the three copies of each bit are correct. In this way, it is sometimes even possible for failures that simultaneously exist in all three buses to be masked.

One of the most interesting characteristics of the C.vmp system is its inherent flexibility and ability to dynamically balance performance and reliability. The ideal result of this characteristic implies a system which can provide a high level of reliability in TMR form at the expense of performance when the situation warrants, and a relatively high level of performance in a loosely-coupled multiprocessor configuration when reliability is not critical.

When in voting mode, the C.vmp system does of course suffer from significant performance degradation. In addition to the obvious loss of two-thirds of potential processor performance, the hardware voting process itself introduces additional propagation delays. However, studies have shown the C.vmp system to exhibit a mean time to failure of between five and six times greater than that provided by a non-redundant LSI-11 processor in the same environment [SIEW78].

2.8 Summary

This chapter has presented a background overview of fault-tolerant computing. A fault-tolerant computer is one that is capable of correct operation despite the presence of certain hardware failures and software errors.

One of the most important principles in any fault-tolerant system is the incorporation of redundancy. The four types of redundancy typically found in fault-tolerant computers are hardware, software, information, and time redundancy. Each is characterized by the addition of extra resources for the purpose of detecting faults, masking faults, tolerating faults, etc.

Another important consideration in any fault-tolerant computer is the choice of communication architecture or interconnection network. Examples include reliable shared buses, shared-memory interconnection networks, loop architectures, tree networks, dynamically reconfigurable networks, binary cube interconnection networks, and graph networks.

The atomic transactions model is used to provide file system update in the presence of faults. This model is taken from magnetic tape transaction processing, and includes such levels of abstraction as careful disk and stable storage.

Communication network properties can be used to model the interconnection network and the processors in a distributed computing system. These properties include no transmission errors, no spontaneous messages, messages delivered in the order sent, and verifiable origination of messages.

Finally, a number of fault-tolerant system examples were presented and discussed. The systems range from commercial computers for fault-tolerant transaction processing to embedded computer systems in space vehicles. Each system incorporates one of more of the basic redundancy approaches and communication architectures.

In order to develop a new system capable of performing DSP operations in a reliable and parallel manner, we will next consider the basic issues of parallel computing followed by an overview of digital signal processing and processors.

Chapter 3
PARALLEL COMPUTING

This chapter provides a background overview of parallel computing, including such issues as parallel processing overview and basic definitions, parallel architecture models, and related background information with regard to operating systems and real-time computing systems.

3.1 Parallel Processing Overview and Definitions

It has been predicted that the decade of the 1990s will usher in the era of true parallel and distributed computing with calculations partitioned into sub-calculations and then given to distinct processors to execute within a multiprocessor system or on a network of processors [DEIT90]. As microprocessors and workstations become more powerful, and the corresponding prices fall, it becomes economically more feasible to incorporate parallel rather than sequential computing and distributed rather than centralized computing. The degree of physical separation of the processors tends to be small due to the fact that communication over long distances is based on telephone lines, which were originally established for highly redundant voice messages at slow speeds. It is instead usually much more efficient to cluster processors at geographically close locations in order to send data and instructions between processors quickly.

As the research areas of parallel processing and parallel computing systems have developed, a number of basic concepts and terms have been introduced. Of these, the following represents some of the fundamental definitions that will be referenced.

The need for parallel processing in future computing systems is becoming increasingly more evident. Non-parallel or sequential computers are quickly approaching the upper limit of their computational potential. This potential performance envelope is dictated by the speed of light, which restricts the maximum signal transmission speed in silicon to 3×10^7 meters per second. Thus, a chip which is 3 centimeters in diameter requires at least 10^{-9} seconds to propagate a signal, thereby restricting such a chip to at most 10^9

floating-point operations per second (i.e. 1 GFLOPS). Since existing supercomputer processors are quickly approaching this limit, the future of sequential processors appears limited, as illustrated in Figure 3.1. And while other chip technologies such as gallium arsenide (GaAs) provide lower signal propagation times than silicon, they only represent a minor delay of the inevitable. Thus, sequential processors are quickly approaching their upper bound, making parallel and distributed computing the wave of the future [DECE89].

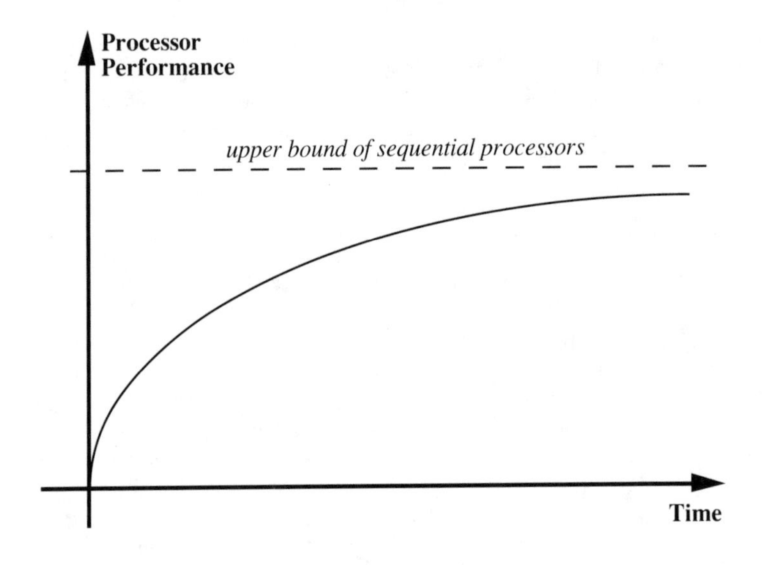

FIGURE 3.1 Trend in Sequential Processor Performance

In its most general form, parallel processing or parallel computing may be thought of as an efficient form of information processing which emphasizes and exploits those events in the algorithm or computing process which are concurrent. The exploitation of these concurrent events may take place at many different processing levels [HWAN84]. The presence of multiple processors in the system is a necessary but not necessarily sufficient condition in order to be considered a parallel system.

The level at which concurrency is exploited can be described in terms of granularity. The *granularity* of a parallel computing system may be defined as the size of the units by which concurrent work is allocated to each of the processors. It is these granularity levels that typically dictate the inherent tradeoff in parallel computers between the size and complexity of the processors and the number of processors. Granularity is often characterized in terms of three levels or grains [DECE89].

Course-grain or large-grain parallel computers involve concurrent processes at the outermost level of program control, such as that attained by decomposing or unfolding the outermost loop of a main program or by executing different jobs on different processors. Course-grain parallel computers typically consist of a small number of large and highly complex processors. Examples include the Cray 2, Cray X-MP, and the Cray Y-MP systems of Cray Inc., each consisting of a few highly sophisticated and powerful processors.

By contrast, *fine-grain* or small-grain parallel computers involve concurrent units of work at the innermost level of execution, such as the individual instructions themselves, the evaluation of expressions, or subsets of each. Fine-grain parallel computers typically consist of a large number of small and simple processors forming what are often referred to as massively parallel processors. Examples include the CM-1 and CM-2 Connection Machines of Thinking Machines Inc., each consisting of tens of thousands of small and simple processors. While fine-grained techniques tend to provide the most potential for performance enhancement, their limitations in terms of algorithm development and communication overhead tends to make them extremely difficult to apply.

Between these two extremes exist the *medium-grain* parallel computers. These systems involve concurrent processes somewhere between the outermost and innermost levels of execution. For example, these units of work might include decomposed inner loops of the main program or subprograms, concurrent inner subroutines, etc. As expected, these machines typically exhibit a balance between the number of processors and the complexity of each processor. Examples include the iPSC/1, iPSC/2, and iPSC/860 systems of Intel Inc., each consisting of dozens of sophisticated and powerful off-the-shelf microprocessors.

Another fundamental issue of parallel computer systems and architectures is the organization and distribution of the system memory. If the memory is distributed or dispersed across the processors in the form of local memories, the system is often called a *loosely-coupled* parallel system. Loosely-coupled systems are often used as *distributed* computers, since they more readily support any requirement to locate the hardware resources of the system in a more physically dispersed or remote manner. However, if the memory is shared by and accessible to all of the processors, in the form of either centralized global memory or shared local memories, the system is often called a *tightly-coupled* parallel system.

Communication between processors in a loosely-coupled parallel system typically takes place using some form of message passing. This form of communication consists of send/receive operations which are carried out between pairs of processors. By contrast, communication between processors in a tightly-coupled parallel system typically takes place using some form of variable sharing in the shared memory system. With variable sharing, one processor can communicate with another by using one or more shared memory locations as the conduit of interaction.

Another fundamental aspect of any parallel computer system is its *interconnection network*. This subsystem is responsible for linking all of the hardware resources in the system, including processors, memory units, and I/O units. Similar to the fault-tolerant communication architectures presented in section 2.4, the forms of these interconnection networks are diverse and numerous. Examples include shared buses, multistage networks, crossbar switches, switch lattices, processor arrays, cubes or hypercubes, and hierarchical connections [DECE89].

The performance of these interconnection networks can be considered in terms of two sometimes conflicting metrics. The first describes the rate at which communications can occur in the interconnection network of the system. This is referred to as the *throughput* or *bandwidth* of the communications subsystem. For example, the rate at which, or the period of time with which, a new output can be produced by a subsystem describes its throughput. By contrast, the *latency* of a communications subsystem describes the total amount of time that is required for a piece of information to travel from the source to the destination.

The fundamental issues of parallel computing tend to form a sequence of considerations and choices. Some of these include course-grain versus fine-grain parallelism, global versus local memory subsystems, centralized versus distributed global system control, variable sharing versus message passing communications, and even general-purpose versus special-purpose systems.

The vast majority of parallel computers developed to date incorporate a control-driven or von Neumann form of execution. With this approach, two or more conventional uniprocessors are connected and execute (either synchronously or asynchronously) traditional fetch-and-execute instruction cycles. The Flynn taxonomy of parallel architecture models is widely used to categorize systems based on the number of instruction streams and data streams managed by the parallel machine [FLYN66]. This taxonomy is presented in the next section.

3.2 Parallel Architecture Models

Parallelism in the architecture of a control-driven machine is obtained by replicating the data stream, the instruction stream, or both [FLYN66]. This presents us with three general types of parallel architectures: single instruction stream with multiple data stream (SIMD); multiple instruction stream with single data stream (MISD); and multiple instruction stream with multiple data stream (MIMD).

The SIMD and MIMD computer organizations have proven to be the more important control-driven parallel architecture types, while MISD has received much less attention and has been challenged as impractical [HWAN84]. The methods by which the SIMD and MIMD parallel architecture types can be subdivided are almost as diverse as the systems themselves. However, DeCegama provides a relatively detailed breakdown in terms of six SIMD models and seven MIMD models [DECE89].

With SIMD machines, all of the processing elements or PEs execute the same instructions at the same time on different elements of the data. SIMD machines can be divided into the categories of: array processors; multiple SIMD machines; pipelined SIMD machines; associative processors, orthogonal processors, and the Connection Machine.

An SIMD array processor consists of a synchronized array of processing elements under the management of a single control unit. Each PE may have its own local memory, access to shared banks of global memory, or both. On these machines, scalar and control instructions are handled by the control unit, while vector and matrix instructions are broadcasted by the control unit to the PEs for computation in a lockstep fashion. As the need arises, individual processors can be disabled or masked in order to address varying problem sizes.

A multiple SIMD machine is simply an array processor which has been enhanced with multiple control units and more PEs so as to form multiple array processors. The control units share the PEs in a dynamic fashion. Pipelined SIMD machines are also an extension to the basic array processor approach. In these machines, the control unit has exclusive access to the global memory modules, and the PEs are configured in a pipeline fashion (i.e. the first PE feeds the second, which feeds the third, etc.).

Both associative and orthogonal processors use associative or content-addressable memory devices so as to perform high-speed read, write, and search functions. Such systems can often perform search and compare functions in parallel by word, in parallel by bit-slice, or both.

Finally, the Connection Machine is essentially an SIMD array processor with features often not associated with such systems. This is a fine-grained SIMD architecture with tens of thousands of processors that can be connected and allocated by software in proportion to the problem size to be addressed. These PE cells can be configured into problem-dependent patterns so that they both represent the structure of the data or algorithm as well as process it. The interconnection network is extensive and includes a large hypercube network as its primary framework.

In the Connection Machine, a host computer sends data through a broadcast network to those nodes that are currently active. A single instruction is then sent to those active processors which is executed in a single machine cycle. Data from this calculation can be sent to neighbors or back to the host. This process is repeated, with single instructions being continually sent to the active processors. This architecture can be used for a variety of tasks such as image analysis, text database searches, VLSI chip design, etc., demonstrating the usefulness of massive parallelism of very simple nodes. The newest version of this system will even employ thousands of advanced RISC microprocessors.

Unlike the SIMD systems, MIMD machines have processors that operate in a relatively independent fashion with respect to each other. With MIMD machines, all of the PEs execute different instructions on different elements of the data. MIMD machines are often called *multiprocessors*, and can be divided into many categories, depending primarily on the interconnection network used. These networks are closely related or in some cases identical to those previously presented in section 2.4. The MIMD categories include: bus systems; multistage systems; crossbar systems; switch lattices; processor arrays; hypercube systems; and hierarchical systems.

As the name suggests, MIMD bus systems consist of independent processors interconnected with each other and perhaps also the memory and I/O units by a common bus or set of buses. While the bus forms a potentially critical source of bottleneck and failure, it represents one of the most straightforward approaches to MIMD processing.

Similarly, MIMD multistage systems consist of independent processors interconnected with each other, and perhaps also the memory and I/O units, by a common multistage network. The switches in the network work together to provide at least one path from any source unit to any destination unit. Traffic can be reduced in these systems by employing switches which are capable of combining similar requests, thereby increasing the overall bandwidth of the system.

Much the same as bus and multistage MIMD systems, the crossbar, hypercube, and hierarchical systems consist of independent units interconnected with each other by a common crossbar switch, binary cube or hypercube network, or hierarchical combinations of various interconnection networks respectively. While the crossbar systems possess the most inherent connectivity between units, they also exhibit the most cost when attempting to increase the number of units. Most others involve inherent tradeoffs between connectivity and cost.

The regularity and connectivity of the generalized binary cube or hypercube architecture makes it an attractive approach for many applications. Another attractive feature of the hypercube is that it can built of hypercubes of lesser dimension. In this way, subcubes of appropriate size can be assigned to a particular problem leaving other subcubes available for other calculations.

Node size varies in existing hypercubes, from the CM-1 and CM-2 Connection Machines with 64K cells to the iPSC/2 series with 8, 16, 32, 64, or 128 80x86 processors. The iPSC/1, the first in Intel's series of hypercubes, was based on 80286 microprocessor nodes together with a cube manager on which programs were developed. Internode communication was provided via a message passing paradigm. The cube manager was in charge of packaging out the parallel tasks. Subsequent iPSC models have used improved communication characteristics, since this plays a major role in the overall approach. This series, as well as NCube Corporation's NCube/Ten Parallel Processing system, is based on the 64-node Cosmic Cube architecture developed at Cal Tech. The basis of these MIMD designs is that, unlike the Connection Machine, each node is an independent processor with its own memory which can communicate with other nodes via message passing. This provides an attractive approach to parallel processing in that it is relatively inexpensive but powerful.

The switch lattice MIMD systems consist of a collection of identical processors placed at regular intervals in a two-dimensional lattice of programmable switches. These systems are typically associated with single-chip VLSI-based multiprocessors using extremely simple processor cells. Each cell consists of a CPU with floating-point capability and local memory. No global memory is provided, and all communication with the outside world is conducted via I/O located at the perimeter of the square lattice. The programmable nature of the switches allows the lattice to be dynamically reconfigured in order to represent the topology of many of the algorithms being implemented.

Finally, MIMD processor arrays or grids consist of loosely-coupled processors which are interconnected in precise patterns with their neighboring processors. These patterns often include linear arrays and meshes as shown in Figures 3.2 and 3.3. Interprocessor communication in an MIMD linear array system takes place with the left or right neighbor of the processor in a bidirectional fashion. Such communication in an MIMD mesh array system takes place with the left, right, upper, or lower neighbor also in a bidirectional fashion. These patterns are often used to exploit the inherent parallelism of the algorithm to be employed.

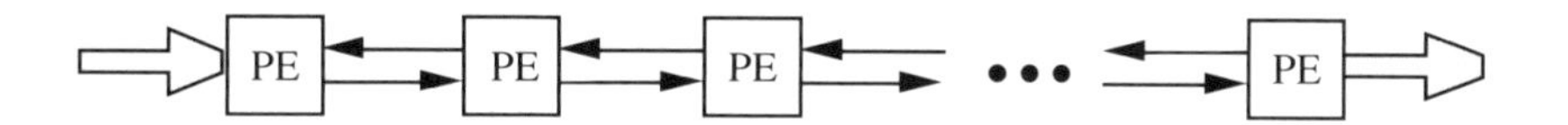

FIGURE 3.2 MIMD Linear Processor Array

In the processor array model, information flows from one cell to its neighbor in a pipeline arrangement with only the boundary cells of the array able to communicate with units not in the array. Applications appropriate for such an architecture are those requiring a regular interconnection structure such as an inner product calculation in large matrix multiplication. The use of parallelism and pipelining increases the execution speed considerably in such cases.

Two basic approaches to processor array implementation are systolic and wavefront arrays. Systems based on the *systolic array* approach are run with a global clock in a synchronous mode, while systems based on the *wavefront array* approach operate asynchronously with a handshaking protocol. Although there is an added time penalty for the handshaking protocol, it is a fixed delay. By contrast, in the systolic array, the problem with clock skewing increases with the size of the array.

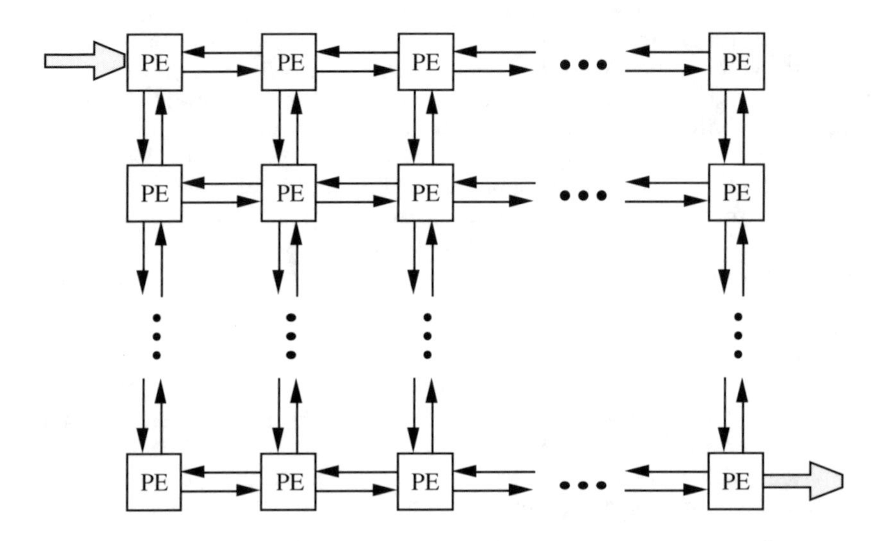

FIGURE 3.3 MIMD Mesh Processor Array

The decision as to which array structure to use for a particular application is made in favor of the systolic array when the processors are simple cells and the number of them is in the order of thousands. In this case, the overhead associated with the handshaking protocol of the wavefront array would be prohibitive. In contrast, the wavefront array is more appropriate when the processors are more complex and/or fault tolerance is desired. This latter capability is difficult to incorporate into the systolic array paradigm as the existence of a faulty processor must be broadcast to all processors and a rollback attempted. In the wavefront array, the processor which is faulty is stopped and subsequent processors can be stopped as a ripple effect through the remaining processors.

In addition to the architecture to be employed, operating systems are also an important aspect of parallel computing and computers. In the next section, an overview of operating systems is presented as they relate to parallel and distributed computing.

3.3 Operating System Overview

A major consideration with any system of parallel or distributed processors is that of the associated operating system. An operating system can be thought of as the programs, implemented in either software or firmware, which make the hardware more usable. These programs form the interface between the hardware and the user, as well as manage the resources of the system [DEIT90].

When considering distributed or higher-grained loosely-coupled parallel computers, an increase in the number of processors brings with it a considerable increase in the number of resources to be managed and allocated efficiently and without conflict. Broadly speaking, the current approaches to this can be put into three classes: a truly distributed operating system; a network operating system; and independent specialized operating systems which allow the processors to communicate via message passing.

Tanenbaum and van Renesse define a distributed operating system as one that appears to the users as a single centralized operating system but that actually consists of independent units (i.e. it is a *virtual uniprocessor*) [TANE85]. Alternately, this can be thought of as a pool of processors that are allocated to tasks as necessary until the tasks are complete, and then the processors are returned to the pool.

By contrast, the network operating system is such that users indeed do know which machine or processor that they are using, and transfers are made explicit by commands. Use of another processor or machine other than the one currently logged into must typically be explicitly accomplished. In this case the processors themselves tend to be fairly sophisticated.

The independent specialized operating systems typically consist of moderately powerful processors, but they are still independent machines with their own scaled down and less sophisticated operating system. Communication between machines is usually accomplished via message passing, with the operating system in charge of the particular handshaking protocols used. While not a complete operating system, these systems can employ a subset of the techniques provided in the more general distributed operating systems in order to address particular application requirements.

The issues associated with the implementation of a distributed operating system can be divided into five categories. These are: communication primitives; fault tolerance; naming and protection; resource management; and services [TANE85].

If the processors in a parallel system do not have access to global memory, basic shared-memory multiprocessor communication techniques such as semaphores do not apply. Instead, some form of message passing must be used to transfer information between processors. While the Open Systems Interconnection (OSI) reference model for message-passing systems provides a versatile approach, the overhead associated with its seven layers is too extensive for a real-time DSP system such as this [ZIMM80]. Instead much simpler techniques are needed.

One of the most widely used models for message passing is the *client-server model,* in which the client processor sends a message to the server with a request, the server services the request, and the server replies back to the client. Three important considerations with simple message passing schemes such as the client-server model are reliable transmissions, blocking, and buffering primitive operations.

In a system with unreliable primitives, when a message is sent, there is no guarantee provided that the message will be received by the appropriate processor or process. The sender is left to either continue on without knowledge of receipt or perhaps explicitly send another message requesting acknowledgement of the first message. In certain applications, reliable primitives may be preferable. In this case, the message sending primitives inherently handle such problems as lost messages, retransmission of messages, and message acknowledgements. These features are provided by the operating system primitives without any direct involvement of the processes involved.

In a system with nonblocking primitives, when a message is sent, control is returned to the sender as soon as the message is queued for transport. Later, after the transmission has completed, the sender is interrupted to indicate that the buffer used for the transmission is now available. In certain applications, blocking primitives may be preferable. In this case, the sender is not allowed to continue execution until the message has completed transmission. While nonblocking provides a higher level of flexibility, it is significantly more complex to implement, since it tends to make programming much more complex and difficult.

Finally, in a system with unbuffered primitives, messages are not buffered. That is, when a message is sent, it is not buffered on the receiving processor so that if the receiver has not executed a receive operation, the sender will be blocked until the receiver does issue a receive request. At that point, the transmission will be completed. This is of course one form of a *rendezvous.* In certain applications, buffered primitives may be preferable. In this case, messages are buffered so that the sender is returned control after a send operation even if the receiver has not issued a receive request.

When a client makes a request of a server, blocks awaiting the service, and regains control when the request is satisfied, this takes on the form of a traditional procedure call mechanism. In fact, this similarity leads to the *remote procedure call* (RPC) model of message passing. This model is particularly interesting from a programming language standpoint, as it gives interprocessor communication similar semantics to that of the local procedure calling mechanism. Some problems that may be associated with this model are concerns over the parameter passing techniques used, how to represent parameters and results in message form, and how to associate particular servers with clients for repetitive accesses.

One of the most important implementation issues of message passing is how to make it efficient. In distributed computing systems, the message passing mechanism itself is often associated with significant bottlenecks. In effect, the overhead associated with the mechanism and the functionality associated with it are at odds. Of course, there is always some fixed amount of overhead associated with any message passing system, in terms of message setup, transmission, reception, and use.

In addition to the communication primitives, other distributed operating system issues are fault tolerance, naming and protection, resource management, and services. With respect to fault tolerance, one advantage of distributed computing systems is that they are often more reliable or available than their centralized counterparts due to the distributed nature of their organization. Naming and protection is concerned with managing process access to objects supported by the operating system such as files, directories, and processes. Resource management is concerned with such issues as processor allocation, process scheduling, load balancing, and deadlocks. Services consist of those functions provided by the operating system to user processes, such as file, print, process, terminal, time, boot, and gateway service.

As pointed out by Tanenbaum and van Renesse, the truly distributed operating system is not easy to implement and few such systems exist [TANE85]. However, some of these mechanisms are relevant and can be employed in more specialized systems in order to address particular application requirements.

3.4 Real-Time Computing Overview

Another area that is related to both parallel and sequential computing is that of real-time computing. Real-time computers are those sequential or parallel systems which are required to acquire data, produce data, or interact with their environment at precise points in time. Such computers are often of critical importance to the industrial and military systems to which they are connected or in which they are embedded. Key issues in such systems include being able to achieve requirements in terms of response time, data acquisition period, etc. [LAWR87].

Although the most recognizable requirement of a real-time computer system is its ability to interact with an industrial process or system in a tightly-coupled manner, another important requirement is often one of reliability, availability, or dependability. Since many systems based on real-time computers are often left unattended and operating without human intervention, it is often necessary that the system be able to suffer a failure without producing a catastrophic result.

Real-time computer systems often communicate with the outside world via sensors and actuators. Sensors are devices which convert a physical signal such as a mechanical movement to an electrical signal such as a voltage change. These electrical signals are usually then conditioned and fed to an

A/D (analog-to-digital) converter for digital input to the real-time computer system. By contrast, actuators are devices which convert electrical signals to physical signals. These devices are usually driven by the conditioned output of a D/A (digital-to-analog) converter which in turn is driven by digital output from the real-time computer system.

Although real-time systems can also be implemented with either analog systems or dedicated non-programmable digital systems, the use of processors in general and microprocessors in particular is often preferable. With a microprocessor-based real-time system, the operations of the system can be conveniently and flexibly expressed and modified in software. Complex functions and parameter changes can be achieved while the system is in operation, and the system can be assembled from off-the-shelf components in a manner which is easily compatible with both analog and digital inputs and outputs.

3.5 Summary

This chapter has presented a background overview of parallel computing and parallel processing. Parallel computing may be thought of as an efficient form of information processing which emphasizes and exploits those events in the algorithm or computing process which are concurrent. The level at which concurrency is exploited can be described in terms of granularity.

The three general types of control-driven parallel architecture are SIMD, MISD, and MIMD. Of these, SIMD and MIMD are the most promising approaches and each general type can be divided into a number of categories. With SIMD machines, all of the processing elements execute the same instructions at the same time but on different elements of the data. By contrast, MIMD machines have processors that operate in a relatively independent fashion with respect to each other, executing different instructions on different elements of the data.

A major consideration with any system of parallel or distributed processors is that of the associated operating system. Broadly speaking, such operating systems include truly distributed operating systems, network operating systems, and specialized operating systems.

Finally, another area that is related to both parallel and sequential computing is that of real-time computing. These computers are required to acquire data, produce data, or interact with their environment at precise points in time and often communicate with the outside world via sensors and actuators. While real-time systems can be constructed using analog or non-programmable digital devices, perhaps the most promising approach is the microprocessor-based real-time system.

Before proceeding with the development of the design for a new system capable of performing DSP operations in a reliable and parallel manner, we will complete the presentation of background overview topics by considering digital signal processing and processors in the next chapter.

Chapter 4
DIGITAL SIGNAL PROCESSING AND PROCESSORS

This chapter provides an overview of DSP and digital signal processors, including such issues as basic algorithms, DSP microprocessors, and the Motorola DSP96002 device.

4.1 Digital Signal Processing Overview

Digital Signal Processing (DSP) is a field rooted in mathematics which has grown to become an important tool in virtually all fields of science and technology. DSP is based on the representation of signals by sequences of numbers and the processing of these sequences. The basic purpose of DSP is to estimate characteristics of a signal or to transform a signal into a more desirable form. In general, signal processing has a rich heritage in a wide variety of disciplines, such as biomedical engineering, acoustics, sonar, radar, seismology, speech recognition, speech synthesis, image processing, and data communication [OPPE75]. It is also closely related to digital control systems.

For example, in applications such as speech recognition, we might be interested in estimating characteristic parameters about the signal, perhaps to verify the identity of the speaker. In applications such as data communication, we might be interested in attenuating part of the signal, perhaps to reduce or eliminate noise and reveal the original information. In addition to one-dimensional signals like these, many applications are based on two-dimensional signals, such as image processing to enhance satellite photographs.

As the name suggests, DSP is concerned with the processing of these signals using digital techniques and devices. Prior to the last few decades, signal processing was typically achieved via analog components. Digital computers were first used in signal processing as a means for simulating their analog counterparts, since digital systems at that point lacked the performance to provide real-time response. However, with the advent of more powerful processors and support devices, the emphasis has changed to digital implementations, due to the inherent advantages of a digital implementation such as flexibility and repeatability. And, as digital systems have continued to provide increasing performance envelopes, more enhanced algorithms have been developed to provide increased signal processing capabilities.

The constant growth in importance that DSP has been experiencing is due to three primary reasons. First, unlike analog components, digital hardware is totally reproducible, so that once a successful piece of hardware has been designed and built, all other manufactured units are guaranteed to perform to exactly the same specifications. Second, there are entire classes of signal analysis and identification problems that can only be solved with a DSP approach. However, as processing requirements have continued to grow, in terms of both speed and complexity, DSP has been limited in its application due to problems in meeting these requirements. Thus, the third reason for the increasing importance of DSP is that improved digital hardware performance, along with improved DSP algorithms, is constantly widening the bandwidth range of signal processing problems that can be solved using DSP [SIMO90].

As noted by Oppenheim and Schafer in 1975, "the importance of digital signal processing appears to be increasing with no visible sign of saturation" [OPPE75]. This conclusion is equally true today. The potential for future capabilities could easily outweigh the strides that have been reached to date. In turn, the proliferation of DSP will continue to promote revolutionary advances in other new fields of application as well.

4.2 Basic Algorithms

The theory and algorithms developed for DSP are widespread and complex. However, most all DSP applications are in some sense related to five primary DSP operations. These operations are digital filtering, convolution, correlation, discrete Fourier transforms via the fast Fourier transform algorithm, and linear algebra operations [ALLE85]. Of these, we can consider digital filters and fast Fourier transforms as representative of potential DSP applications. Both of these operations will be overviewed and described in terms pertinent to their implementation on a digital computer.

4.2.1 Digital filters

A filter is generally a device which selects certain frequencies of the input signal to pass and certain ones to attenuate or block [JOHN89b]. Which parts of a signal are passed and which are attenuated is determined by the system transfer function, an equation in the frequency domain. Thus, digital filters are described in the frequency domain by transfer functions and in the time domain by difference equations (DEs). Filters are often categorized in terms of their action upon frequencies. For example, *low-pass* filters will pass low frequencies and attenuate high frequencies, and *high-pass* filters do just the opposite. *Band-pass* filters attenuate all but a specific range of frequencies, and *band-stop* filters do just the opposite.

Most filtering applications are based on systems that are linear and time-invariant (LTI). The class of LTI digital filters is composed of two basic types. They are infinite impulse response (IIR) or recursive digital filters, and finite impulse response (FIR) or nonrecursive digital filters.

Recursive or IIR digital filters are described by difference equations whose outputs are a function of the current and previous inputs along with previous outputs. By contrast, nonrecursive or FIR digital filters are described by DEs whose outputs are a function of *only* the current and previous inputs, and not previous outputs.

By using algebraic manipulation, the form of the defining DE of a digital filter can be altered so as to modify the overall algebraic form and computational requirements of its implementation. These DEs are commonly implemented by dividing them into second-order sections in either a cascaded (via factoring of the transfer function) or a parallel (via partial fraction expansion of the transfer function) fashion. One common IIR formulation, shown below, is referred to as 3D in [PHIL84] and Direct Form I in [BOWE82], where $x(k)$ is the input, $y(k)$ the output, a_i and b_i the coefficients, and n the order of the filter:

$$y(k) = a_0 x(k) + a_1 x(k-1) + \ldots + a_n x(k-n) - b_1 y(k-1) - \ldots - b_n y(k-n)$$

$$\Rightarrow y(k) = \sum_{i=0}^{n} a_i x(k-i) - \sum_{i=1}^{n} b_i y(k-i)$$

$$\Rightarrow y(k) = a_0 x(k) + \sum_{i=1}^{n} \{a_i x(k-i) - b_i y(k-i)\} \quad (4.1)$$

This form can readily be adapted to the FIR case by making all of the b_i coefficients equal zero. In either case, clearly the determining factor in implementing digital filters is the ability to handle multiplications and additions, along with data transmission, at a very high rate. The faster the processing rates, the higher the achievable sample frequencies, and thus the larger the signal bandwidth that may be filtered. In addition to the maximum sampling rate, other important issues in the implementation of a digital filter are internal precision, input/output precision, and the ratio of sampling to signal frequencies. The precision effects can include coefficient quantization, input/output variable quantization, and product quantization noise [PHIL84].

4.2.2 Fast Fourier transforms

Discrete Fourier Transforms (DFTs) are widely used in DSP applications to characterize signals in terms of their spectral frequency content. DFTs are commonly implemented using the Fast Fourier Transform (FFT) algorithm, since while the discrete Fourier transform is an $O(n^2)$ algorithm, the equivalent FFT is an $O(n\log_2 n)$ one. The FFT and its inverse provide a mechanism for converting signals from the time domain to the frequency domain and back respectively. For example, a signal may be designed and described via a frequency spectrum, and converted using the inverse FFT to the time domain for real-time generation. Similarly, a signal may be modified in the time domain and converted using the FFT to the frequency domain for analysis.

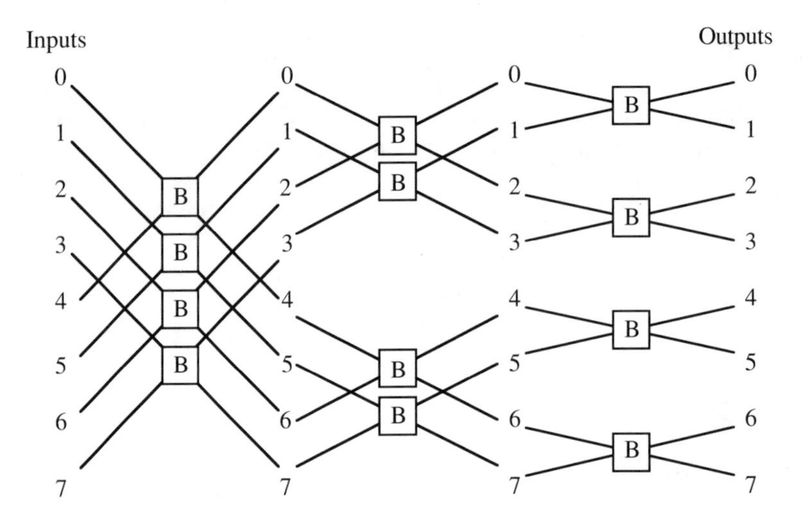

FIGURE 4.1 An Example on an 8-point FFT Decomposition

Each FFT is defined in terms of N input/output complex or real data points. The FFT algorithm decomposes the problem into a number of computations known as *butterflies* [ALLE85]. It is these butterflies that form the fundamental principle of the FFT algorithm and its efficiency. The graphical representation of an N-point FFT consists of $\log_2 N$ stages of $N/2$ butterflies each. For example, an 8-point FFT consists of three stages of four butterflies each, for a total of twelve butterflies (see Figure 4.1).

As the example illustrates, the inputs to the first stage of butterflies are separated by $N/2$ points, the inputs to the second stage are separated by $N/4$ points, etc. In general, each butterfly is identical in form, requiring one complex multiply and two complex adds. This can be realized via four real multiplies and two real adds for the complex multiply, and two real adds for each complex add, for a grand total of four real multiplies and six real adds for each butterfly. Thus, it is evident that, like digital filtering, the determining factor in implementing an FFT is the ability to handle multiplications and additions, along with data transmission, at a very high rate.

It is clear from the graphical description of the FFT that there exists much potential for parallel implementation of the FFT. For instance, for an N-point FFT where $\log_2 N$ identical processors are available, each could be used for one stage in a pipeline or linear array fashion. In this way, each processor calculates $N/2$ butterflies, thereby making the overall response equal to the time it takes a processor to perform $N/2$ butterflies and transfer its results once the linear array has filled. Similarly, if $N/2$ processors are available, each of the processors can be dedicated for one of the butterflies in each stage, thereby making the response time equal to the time it takes a processor to perform $\log_2 N$ butterflies and transfer its results. If only a few processors are available, an even simpler approach would be to apply all processors to part of each butterfly, taking advantage of the inherent parallelism in butterfly calculations [ALLE85].

Of course, the limitations to these types of approaches are the number of data points N, the number of processors, and the interconnection network used to pass information between processors. For example, a 1024-point FFT implemented using one processor per stage would require 10 processors, each of which being responsible for the computation of 512 butterflies. Also, it should be noted that while we have described the FFT in terms of base two, other larger bases are sometimes used.

4.3 Digital Signal Processing Microprocessors

One of the most important recent developments in scientific computing in general and DSP in particular is the DSP microprocessor. These devices are designed to perform numerically intensive real-time DSP tasks at an optimum rate. They typically consist of a single chip often supplemented with support chips for I/O, A/D conversion, D/A conversion, and additional off-chip memory. While designed with performance in mind, they are considered weaker in functionality and programming convenience than their general-purpose microprocessor counterparts. This situation, however, is rapidly changing [LEE88, LEE89].

The development of DSP microprocessors began in the late 1970s, via companies including Intel, Bell Labs, and NEC. Since that time, companies such as Texas Instruments, Motorola, Hitachi, and Fujitsu have focused a large part of their microprocessor strategies around DSP applications. Today, programmable DSP devices have become a highly lucrative and competitive industry.

In order to achieve the performance improvements typical of DSP microprocessors, a number of architectural strategies have been employed. Three of the most important techniques that are often employed are specialized arithmetic hardware for high-speed operations, multiple buses for parallel memory access, and pipelining for parallel execution.

4.3.1 Specialized arithmetic hardware

Perhaps the most fundamental characteristic of DSP microprocessors is their incorporation and integration of high-performance hardware multipliers and adders. Many devices are capable of performing a multiply-accumulate (i.e. a simultaneous multiply and addition) in a single instruction cycle. Typical multiply-accumulate times can range anywhere in the tens to hundreds of nanoseconds range, depending on the device.

DSP microprocessors can be categorized by the method used for arithmetic storage and manipulation, either fixed-point or floating-point. While fixed-point systems are simpler to design, they exhibit inherent difficulties with scaling. As increasing numbers of multiplies and additions are performed, the size of the result tends to exceed the limitations of the fixed-point scheme. Thus, the results must be scaled in order to be used in subsequent operations and circumvent overflow. By contrast, floating-point systems avoid these limitations, since numbers are represented by both a mantissa and an exponent. Thus, in the vast majority of cases, the results of an arithmetic operation is still representable in the same format without modification or consideration of overflow. In fact, most devices provide hardware to automatically re-normalize the resulting floating-point values. However, the disadvantage of floating-point systems is the cost and complexity required to support floating-point numbers as the base representation, although this is diminishing.

For example, in terms of arithmetic significance, consider two systems with numbers represented in n bits, one fixed-point and one floating-point. With fixed-point numbers, more fractional precision is provided as long as values fall within a rigid range of values (e.g. only k bits of integer and n-k bits of fractional data). Values outside this range cannot be represented directly, but instead are scaled into this range. With floating-point numbers, the n bit value is split into m bits for the mantissa and n-m bits for the exponent. While the precision of the value itself is limited by the size of the mantissa, the exponent provides a much wider range of possible values. Thus the need for scaling is virtually eliminated.

4.3.2 Multiple buses

The vast majority of conventional general-purpose microprocessors share a common trait with all "von Neumann" architectures. They incorporate a single set of buses for communication between the microprocessor and the memory and I/O devices. This set of buses consists of an address bus, to indicate the address to be accessed, the control bus, to indicate the type of access (e.g. read or write), and the data bus, upon which the actual data to be transferred is relayed. While this approach provides a simple and effective mechanism for microprocessor and memory communication, it does form a bottleneck whereby only one memory access can occur at any point in time. For example, in a typical two-operand instruction cycle, a memory access takes place to fetch the opcode, another to fetch the first operand, another to fetch the second operand, and a final memory access during execution of the instruction itself. Since memory access is relatively slow compared to internal operation in a microprocessor, a severe bottleneck is identified.

A natural evolution for these devices is to instead make use of multiple sets of buses, coupled with multiple memory devices, thereby allowing more than one memory access to take place simultaneously. The use of multiple buses for parallel memory access is widespread throughout virtually all DSP microprocessors. The reasons for using this method are two-fold. First, DSP applications typically require computational performance at the cutting edge of technology. Thus, every method must be considered to increase numeric computational performance. Second, unlike some applications, DSP problems are often easily partitioned into two or more distinct address spaces. For example, when implementing a digital filter, the coefficients, the input and output delay variables, and the program itself are all distinct entities. Thus, by providing three separate address spaces, all three types of data can be potentially accessed in parallel.

The number and type of multiple buses used in DSP microprocessors varies from device to device. For example, early devices incorporated a basic Harvard architecture, where program and data information is separated into distinct address spaces. Most recent devices extend this architecture by either providing more than two address spaces, providing a program cache, or a combination of both. For example, the Motorola DSP96002 provides three address spaces (a program space, an X data space, and a Y data space) which can be accessed concurrently [DSP89].

Another key issue for DSP microprocessors is internal versus external memory. Many devices provide on-chip memory in a fashion similar to conventional microcontrollers. However, the problem with external memory is the performance penalties often associated with them. That is, some DSP microprocessors exhibit single-cycle memory access only for internal memory, and are much slower for external memory.

4.3.3 Pipelining

Closely related to the multiple buses provided in DSP microprocessors, most devices have an architecture centered around a pipelined approach to parallel processing. That is, like an automobile assembly line, different parts of different instructions are executed by separate stages of the processor independently and simultaneously. Once the pipeline has been filled, and while it stays filled, overall results can be produced at the rate of a single stage in the pipeline.

For example, while one instruction is experiencing an opcode fetch, the previous instruction might be experiencing its operand fetches, and the instruction prior to that its execution. Thus, pipelining provides a means by which all of the parallel buses can be kept busy while parallel processing takes place in the processor via concurrent arithmetic operations. This is necessary, as it is typically not possible for a single instruction to experience an opcode fetch, operand fetches, and execution simultaneously, since the operands depend on the opcode and the execution depends on both the opcode and operands (i.e. the parts of the instruction are interdependent).

The techniques used by DSP microprocessors to facilitate pipelining can be divided into three categories: interlocking, time-stationary coding, and data-stationary coding. Each is characterized in terms of how the programmer describes the pipelined parallel execution of instructions [LEE88, LEE89].

With the interlocking approach, the programmer is not bothered with the details of pipelining. That is, the programmer writes code without considering whether or not the previous instructions have completed execution. If and when a conflict occurs, such as an instruction requiring the results of a previous instruction that has yet to complete execution through the pipeline, the control hardware delays or *interlocks* the execution of the instruction. Of course, in this case, performance is degraded. In addition, it can be difficult to anticipate exactly how many cycles an instruction will require, since it will depend on neighboring instructions.

With the time-stationary coding approach, the programmer is given more explicit control over the stages of the pipeline. This is typically accomplished by having each instruction specify all the operations that are to occur simultaneously in a single instruction cycle. For example, an instruction might indicate that two operand fetches, two pointer register updates, a multiply and an accumulate are to take place concurrently. In effect, the program model is one of simultaneous as opposed to successive execution. When conflicts in access appear, such as a register being used as the destination of an operand fetch and as a source for a multiply-accumulate operation, the source value is in effect taken as that resulting from the *previous* instruction affecting that register. That is, while the new value of the register is being calculated, the old value is being used, thereby avoiding the conflict.

Finally, the data-stationary coding approach incorporates instructions that specify all of the operations to be performed on a single set of operands from memory. Instead of specifying what happens to the hardware at a particular point in time, instructions using this approach specify what happens to the data throughout its processing. Thus, the results of each instruction are often not immediately available for subsequent instructions.

One problem with pipelining that has been well documented is the handling of conditional branches. These branches make it difficult to keep the pipeline operating at peak efficiency, since at any moment a jump to a different location may be required, which in turn dictates the need to perhaps flush the contents of the entire pipeline. Techniques have been developed to reduce the penalties of these branches, such as those used in RISC (Reduced Instruction Set Computer) microprocessors [PATT85]. However, many DSP microprocessors instead avoid this problem by incorporating specialized low-overhead looping instructions for tight inner loops instead of conditional branch instructions.

4.4 The DSP96002 Digital Signal Processor

4.4.1 Introduction

The DSP96002 is the newest member to the Motorola line of digital signal processors. The 40 MHz version of this CMOS chip makes use of a high degree of internal parallelism to provide a peak performance of approximately 60 MFLOPS, and is ideally suited for all types of DSP applications. We will overview the principal features of this new microprocessor. Complete and detailed descriptions of all of the DSP96002 hardware and software features can be found in [DSP89].

The DSP96002 is the first 32-bit DSP microprocessor to implement in hardware the IEEE 754-1985 standard for binary floating-point arithmetic [IEEE85]. It features an extended Harvard architecture, where both a 4 GW (gigaword) program address space and two 4 GW data address spaces (X and Y) are supported, as well as a highly-optimized pipelined program control unit. Two complete sets of 32-bit buses or *ports* are provided for external interfacing. Similar to a conventional microcontroller, the DSP96002 provides a number of on-chip memory features. These include 1 KW (kiloword) of program RAM, 1 KW of data RAM (512 words each for X and Y space), 1 KW of data ROM (512 words each for X and Y space), a dual address generation unit, and a dual-channel DMA (direct memory access) controller (see Figure 4.2) [KLOK89a].

Parallel data movement is provided by eight 32-bit internal buses. These are the XAB (X address bus), YAB (Y address bus), PAB (program address bus), XDB (X data bus), YDB (Y data bus), PDB (program data bus), GDB (global data bus), and DDB (DMA data bus). The address buses provide dual-access for simultaneous DMA and core transfers. In this way, four 32-bit transfers (instruction prefetch, two parallel data moves, and one DMA transfer) can occur during each instruction cycle.

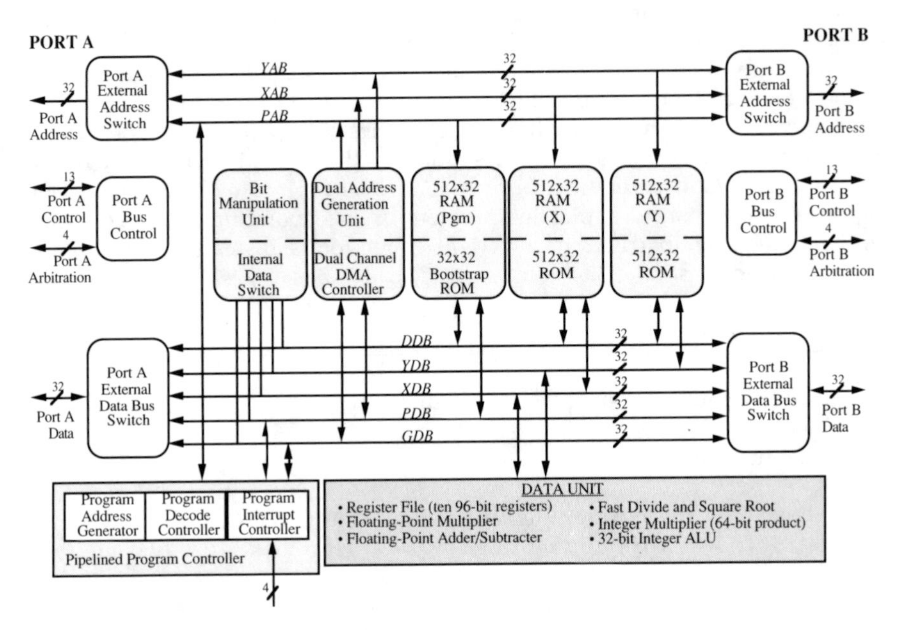

FIGURE 4.2 Motorola DSP96002 Block Diagram

Floating-point operations are provided by a hardware multiplier and an add/subtract unit. Both 32-bit single precision (SP) and 44-bit single extended precision (SEP) floating-point formats of the IEEE 754 standard are supported. In addition, the DSP96002 is upward compatible with its fixed-point predecessor, the DSP56001, supporting all fixed-point operations via internal integer arithmetic hardware.

The DSP96002 uses a pipelined program controller which oversees all instructions, where one instruction cycle is equivalent to two clock cycles. Most instructions occupy one word of memory and execute in one instruction cycle. The execution unit is register-oriented, in that all arithmetic instructions are with respect to the floating-point register file. This file is reconfigurable as either ten 96-bit or thirty 32-bit floating-point registers. The parallel hardware and instruction set support simultaneous arithmetic operations and data transfers, allowing up to three floating-point operations (a multiply and add/subtract combination), two parallel data memory moves, and two pointer register updates all in a single instruction cycle.

The DSP96002 uses the time-stationary coding model of pipelined parallel instruction execution. With this model, the programmer is given explicit control over the instructions that are to occur concurrently in a single instruction cycle and the associated pipeline stages. For example, an FP multiply and add with two parallel data moves would use an instruction like the following:

```
FMPY D2,D3,D0     FADD D0,D1     X:(R0)+,D4     Y:(R4)+,D5
```

Here the contents of registers D2 and D3 are multiplied and stored in D0. At the same time, the previous contents of D0 (as it appeared prior to this instruction) is added to D1, while new values are loaded into D4 and D5 for use by the next instruction.

In order to address the problem associated with conditional branches and pipelining, highly developed low-overhead loop instructions are provided. These loop instructions avoid the inefficiencies of conditional branching techniques by using a predetermined loop-count to define tight inner loops of any length with a minimal setup time.

Many DSP operations require special means for accessing data in memory. For example, digital filters need circular data arrays or buffers in order to delay input and output values, while FFTs need bit-reversing capability in their data arrays. The DSP96002 supports these accessing requirements in hardware via built-in addressing modes.

A dedicated serial port is provided for on-chip circuit emulation, enabling interactive test and debug of the chip during operation without dedicated monitor code or cabling. One of the configuration modes of the DSP96002 allows the processor to bootstrap over one of the external ports. In addition, extra circuitry is provided with each port for bus arbitration in shared-bus multiprocessor configurations, as well as interfacing support for DRAM and VRAM memory devices. System development is supported by a cross-assembler, a linker, and a callable multi-device simulator subroutine library with single-cycle granularity and complete interface simulation.

4.4.2 Programming model

Programs for the DSP96002 microprocessor can be viewed as instructions that control three execution units operating in parallel. These execution units are the data ALU unit, the address generation unit, and the program controller unit. An extensive and versatile set of registers is provided for use by the instructions in order to feed these units and keep each as busy as possible [DSP89].

The registers can be grouped into the following categories based on their function:

- user data registers
- user address registers
- user offset registers

- user modifier registers
- system support registers

The user data registers are used by the data ALU unit, while the user address registers, user offset registers, and user modifier registers feed the address generation unit. The system support registers are used by the program controller unit to determine the control-flow and status of the programs being executed. A diagram of the programming model is shown in Figure 4.3.

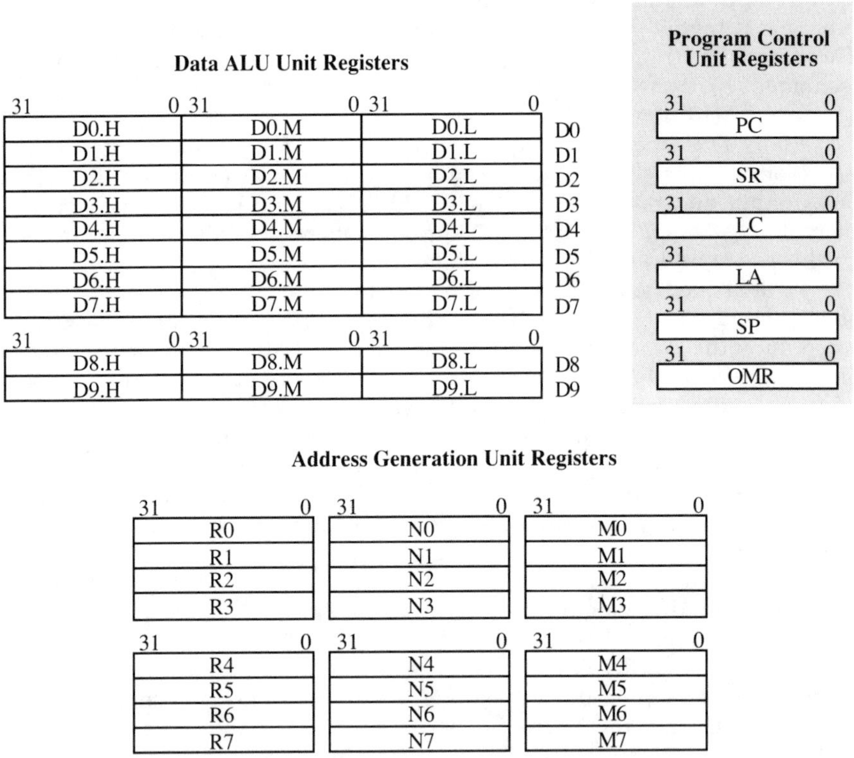

FIGURE 4.3 DSP96002 Programming Model

The user data registers consist of ten registers, D0-D9, which are each 96-bits wide. Each of these registers can be treated as a single 96-bit register or three independent 32-bit registers. When divided into three registers, special designators which indicate low, medium, and high word are used. For example, the three 32-bit registers associated with D4 can be individually referenced by using their word designator via:

D4.L	{ for the low word associated with D4 }
D4.M	{ for the middle word associated with D4 }
D4.H	{ for the high word associated with D4 }

Fixed-point arithmetic operations are performed using the low word data registers D0.L-D9.L. Floating-point arithmetic operations use the 96-bit registers D0-D9 in an internal representation of the IEEE double-precision format. Of these ten registers, D8 and D9 are primarily included for support of four-instruction radix-2 FFT butterflies. Unlike D0-D7, operations with D8 and D9 are limited to source operands in multiply operations and source or destination operands in MOVE instructions.

A set of eight 32-bit user address registers are used to hold memory pointers or indexes as well as general-purpose data. These eight registers, R0-R7, are split into register files for accessing any two of the three address spaces. The first set is made up of registers R0-R3 which can be used for pointing to memory locations in any one of the address spaces. Similarly, the second set consists of registers R4-R7 which can be used to point into either of the other two address spaces. For example, a register from the first set might point into the X space while at the same time a register from the second set points into the Y space. The address contained in any of the eight address registers may point directly to the desired data in memory, or it may be modified by a user offset register.

In cooperation with the address registers, a set of eight 32-bit user offset registers is provided for address register update calculations as well as general-purpose storage. As with the address registers, these eight registers, N0-N7, are split into register files for accessing any two of the three address spaces. The first set is made up of registers N0-N3 which can be used for incrementing or decrementing address register pointers into one of the three address spaces. Similarly, the second set consists of registers N4-N7 which can be used for incrementing or decrementing address register pointers into either of the other two address spaces. In either case, when used for addressing, an offset register is combined with the address register of the same number in order to produce an address from the sum of their contents. For example, N0 is used as an offset from the address pointed to by R0, N1 is used with R1, etc.

Along with the address and offset registers, a set of eight 32-bit user modifier registers is provided to implicitly specify address arithmetic types to be used in address register update calculations as well as for general-purpose storage. As before, these eight registers, M0-M7, are split into register files made up of M0-M3 and M4-M7 for independent address space access. Any one of three access specification types can be stored in a modifier register: linear; reverse carry; and modulo. With a linear modifier, the address access is performed using normal 32-bit linear arithmetic. The reverse carry modifier is used to support the bit-reversing address access often associated with FFT algorithms. Finally, the modulo modifier is used to access circular buffers in memory for such applications as delayed inputs and outputs in IIR digital filters. As expected, M0 designates the access type of R0 with or without N0, M1 designates the access type of R1 with or without N1, etc.

The final set of registers provided in the programming model of the DSP96002 microprocessor consists of assorted system support registers. A 32-bit program counter register (PC) is used to hold the address of the next location to be fetched from the program address space. A 32-bit status register (SR) is used to hold numerous condition codes and flags. A 32-bit loop counter register (LC) and a 32-bit loop address register (LA) are used together to support high-speed hardware looping by holding the number of iterations to be performed by the loop and the address of the last instruction word in the program loop respectively. A 32-bit stack pointer register (SP) is used to indicate the location of the top of the system stack. This stack consists of fifteen 64-bit locations in separate internal RAM and is used for storing such items as return addresses from subroutine and interrupt calls as well as the address of the beginning instruction for a hardware program loop. Finally, a 32-bit operating mode register (OMR) is used to define the current chip operating mode of the processor. The bits in the OMR are used to select whether or not internal program RAM is to be enabled, what the startup procedure or bootstrap mode of the chip is to be, whether the sine/cosine internal data ROMs are to be enabled, etc.

4.4.3 Addressing modes

A number of operand addressing modes are provided for use by the DSP96002 instruction set. These modes can be divided into four general categories: register direct modes; address register indirect modes; PC relative modes; and special addressing modes.

The register direct addressing modes consist of data or control register direct and address register direct. With data or control register direct, the operand is in one, two, or three data ALU unit register(s) or program controller unit register(s) as indicated in the data movement field of the instruction. With address register direct, the operand is in one of the twenty-four address generation unit registers as specified in the instruction. Both modes are classified as register references. An example for each of these register direct modes is shown below:

```
FCLR  D2            ; clear floating-point register D2 (data register direct via D2)

MOVE  #0, N4        ; move 0 to offset register R4 (address register direct via N4)
```

The address register indirect modes consist of eight mechanisms for which the contents of any address register can be used to determine the address to be accessed in memory. These include: indirect with no update; indirect with post-increment by one; indirect with post-decrement by one; indirect with post-increment by an offset; indirect with post-decrement by an offset; indirect indexed by an offset; pre-decrement and then indirect; and indirect with long displacement.

In all cases, the address register serves as a pointer into one of the three address spaces and the type of address access is implicitly and automatically specified by the contents of the respective modifier register. The actual address to be accessed in this indirect fashion is calculated by the address generation unit. An example for each of these eight address register indirect modes is provided below in the same order as introduced:

```
MOVE X:(R0), D3.s      ; move the single-precision X space value pointed to by R0 to D3
                       ; without changing the contents of R0

MOVE X:(R0)+, D3.s     ; move the single-precision X space value pointed to by R0 to D3
                       ; and then increment the contents of R0

MOVE X:(R0)-, D3.s     ; move the single-precision X space value pointed to by R0 to D3
                       ; and then decrement the contents of R0

MOVE X:(R0)+N0, D3.s   ; move the single-precision X space value pointed to by R0 to D3
                       ; and then add the signed contents of N0 to R0

MOVE X:(R0)-N0, D3.s   ; move the single-precision X space value pointed to by R0 to D3
                       ; and then subtract the signed contents of N0 from R0

MOVE X:(R0+N0), D3.s   ; move the single-precision X value pointed to by (R0+N0) to D3
                       ; without changing the contents of R0 or N0

MOVE X:-(R0), D3.s     ; decrement the contents of R0 and then move the single-
                       ; precision X space value now pointed to by R0 to D3

MOVE X:(R0+LB), D3.s   ; move the single-precision X value pointed to by (R0+LB) to D3,
                       ; where LB is a one word label signifying a memory address
```

Due to the pipelined instruction processing provided by the DSP96002 to increase throughput, if any of the address generation unit registers are updated with a MOVE instruction, the new contents of that register will not be available for use as a pointer until the second instruction after the MOVE is executed.

The PC relative addressing modes consist of mechanisms to add a displacement to the contents of the PC in order to determine a new value for the

PC. These modes include long displacement, short displacement, and displacement via an address register.

Finally, the special addressing modes specify the operand or address of the operand in a field of the instruction or they reference an operand in an implicit fashion. These modes include immediate data, immediate short data, absolute address, absolute short address, short jump address, I/O short address, and implicit reference.

4.4.4 Instruction set

Due to the parallel nature of the buses and hardware units in the DSP96002, as many as three data transfers can be explicitly specified in an instruction word. These include transfers on the X data bus, the Y data bus, and within the data ALU unit. In addition, a fourth implicitly defined data transfer is generally involved in the instruction processing itself and occurs in the program controller unit (e.g. instruction fetches). Each transfer is specified by a source and a destination operand.

Most instructions used on the DSP96002 consist of one or more data ALU operations, an X space data transfer, and a Y space data transfer all in the same instruction. Assembly language source code for the DSP96002 uses the following general format (where items in brackets are optional):

```
[label]   opcode operand(s)  [opcode operand(s)]  [X-xfer]  [Y-xfer]   [; comments]
```

The first opcode field specifies the data ALU, address generation, or program controller unit operation to be performed as well as operations for the bit-manipulation unit. The operand(s) for this operation are indicated immediately following the opcode. The second opcode field specifies an optional floating-point adder/subtracter operation for the data ALU unit which, along with the first opcode, provides support for parallel operation of the floating-point multiplier and the adder/subtracter. The operand(s) for this second operation are also indicated immediately after the opcode.

The next field in a general DSP96002 assembly language instruction is used to specify optional data transfers over the X bus along with the source and destination addressing modes for the operands to be used in the transfer. Similarly, the next field is used to specify optional data transfers over the Y bus along with the operands and their addressing modes. In both fields, the transfers take place in parallel with the data ALU operations specified in the opcodes. Address space qualifiers such as X: and Y: are used in these two fields to indicate which data address space is to be referenced. For example:

```
opcode operands     opcode   operands   X-xfer         Y-xfer
FMPY  D0,D5,D2      FSUB.s   D7,D3      X:(R0)+,D0.s   Y:(R4)+,D5.s
```

The execution of instructions on the DSP96002 is carried out using pipelining so that a peak throughput of one instruction per instruction cycle can be achieved. However, certain instructions require an additional instruction cycle to execute, such as instructions that are longer than one word and instructions that cause a change in the control flow.

The DSP96002 instruction set consists of 133 instructions which can be divided into seven groups. These include: 38 floating-point arithmetic instructions; 30 fixed-point arithmetic instructions; 13 logical instructions; 4 bit-manipulation instructions; 4 loop instructions; 9 move instructions; and 35 program control instructions.

Each of the 38 floating-point arithmetic instructions operates using one or more of the 96-bit data ALU unit registers. The DSP96002 supports operations on both IEEE standard single-precision (8-bit exponent and 24-bit mantissa) and IEEE standard single-extended-precision (11-bit exponent and 32-bit mantissa) operands. Some of these instructions use both of the opcode and operand(s) fields in order to support multiplication, addition, and subtraction in parallel. Since the floating-point instructions are register-based and execute within the data ALU unit, optional parallel data transfers are supported in the X and Y fields. A brief summary for each of these instructions is provided below:

```
FABS.s        ; single-precision absolute value
FABS.x        ; single-extended-precision absolute value
FADD.s        ; single-precision add
FADD.x        ; single-extended-precision add
FADDSUB.s     ; single-precision add and subtract
FADDSUB.x     ; single-extended-precision add and subtract
FCLR          ; operand clear
FCMP          ; comparison
FCMPG         ; graphics comparison
FCMPM         ; magnitude comparison
FCOPYS.s      ; single-precision sign copy
FCOPYS.x      ; single-extended-precision sign copy
FGETMAN       ; get mantissa
FINT          ; convert to floating-point integer (rounded)
FLOAT.s       ; single-precision integer->float
FLOAT.x       ; single-extended-precision  integer->float
FLOATU.s      ; single-precision unsigned integer->float
FLOATU.x      ; single-extended-precision unsigned integer->float
```

```
FLOOR              ; convert to floating-point integer (truncated)
FMPY  FADD.s       ; single-precision multiply and add
FMPY  FADD.x       ; single-extended-precision multiply and add
FMPY  FADDSUB.s    ; single-precision multiply, add, and subtract
FMPY  FADDSUB.x    ; single-extended-precision multiply, add, and subtract
FMPY  FSUB.s       ; single-precision multiply and subtract
FMPY  FSUB.x       ; single-extended-precision multiply and subtract
FMPY.s             ; single-precision multiply
FMPY.x             ; single-extended-precision multiply
FNEG.s             ; single-precision negation
FNEG.x             ; single-extended-precision negation
FSCALE.s           ; single-precision scaling
FSCALE.x           ; single-extended-precision scaling
FSEEDD             ; reciprocal approximation
FSEEDR             ; square-root reciprocal approximation
FSUB.s             ; single-precision subtract
FSUB.x             ; single-extended-precision subtract
FTFR.s             ; single-precision register transfer
FTFR.x             ; single-extended-precision register transfer
FTST               ; operand test
```

As with the floating-point instructions, each of the 30 fixed-point arithmetic instructions operates within the data ALU unit and supports parallel data transfers in the X and Y fields. Both signed and unsigned 32-bit fixed-point operands are supported. A brief summary for each of these instructions is provided below:

```
ABS                ; absolute value
ADD                ; add
ADDC               ; add with carry
ASL                ; arithmetic shift left
ASR                ; arithmetic shift right
CLR                ; operand clear
CMP                ; comparison
CMPG               ; graphics comparison
DEC                ; decrement by one
```

```
EXT        ; sign extend from 16-bits to 32-bits
EXTB       ; sign extend from 8-bits to 32-bits
GETEXP     ; get exponent
INC        ; increment by one
INT        ; signed float->integer conversion
INTRZ      ; signed float->integer conversion with round to zero
INTU       ; unsigned float->integer conversion
INTURZ     ; unsigned float->integer conversion with round to zero
JOIN       ; join two 16-bit integers
JOINB      ; join two 8-bit integers
MPYS       ; signed multiply
MPYU       ; unsigned multiply
NEG        ; negate
NEGC       ; negate with carry
SETW       ; set an operand to all binary ones
SPLIT      ; extract a 16-bit integer
SPLITB     ; extract an 8-bit integer
SUB        ; subtract
SUBC       ; subtract with carry (borrow)
TFR        ; transfer 32-bit data ALU unit register
TST        ; test an operand
```

With the exception of the ANDI and ORI instructions, each of the 13 logical instructions execute within the data ALU unit. As with the floating-point and integer arithmetic instructions, logical instructions (except ANDI and ORI) are register-based so that optional parallel data transfers can be specified in the X and Y fields. A brief summary for each of these instructions is provided below:

```
AND        ; logical AND
ANDC       ; logical AND with complement
ANDI       ; AND immediate value to control register
BFIND      ; find leading binary one
EOR        ; logical XOR (exclusive-OR)
LSL        ; logical shift left
LSR        ; logical shift right
NOT        ; logical complement
```

OR	; logical OR (inclusive-OR)
ORC	; logical OR with complement
ORI	; OR immediate value to control register
ROL	; rotate left
ROR	; rotate right

The bit-manipulation instructions are used to test and optionally modify any single bit in a data memory location or register. The result of the bit test is implicitly stored in the carry bit of the status register SR. Unlike most of the previous instructions summarized thus far, bit-manipulation instructions *do not* support parallel data transfers with the X and Y fields. A brief summary for each of these instructions is provided below:

BCLR	; bit test and clear
BSET	; bit test and set
BCHG	; bit test and change (flip)
BTST	; bit test

The loop instructions are used to control the hardware looping facility in the DSP96002 by starting a loop, preparing parameters for a loop, and cleaning up after a loop. A brief summary for each of these instructions is provided below:

DO	; start a hardware loop
DOR	; start a PC relative hardware loop
ENDDO	; exit from a hardware loop
REP	; repeat the next instruction a specified number of times

The move instructions are used to perform data transfers over the data buses in addition to the those provided in parallel with the arithmetic and logical instructions previously described. A brief summary for each of these instructions is provided below:

LEA	; load effective address
LRA	; load PC relative address
MOVE	; move data register(s)
MOVETA	; move data register(s) and perform an address calculation test
MOVEC	; move control register

```
MOVEI      ; move immediate
MOVEM      ; move program memory
MOVEP      ; move I/O peripheral data
MOVES      ; move absolute short
```

The 35 program control instructions on the DSP96002 are used to modify the control-flow of the program control unit. A brief summary for each of these instructions is provided below (where 'cc' denotes a reference to one of the condition codes in the status register SR):

```
Bcc        ; conditional branch
BRA        ; unconditional branch
BRCLR      ; branch if bit clear
BRSET      ; branch if bit set
BScc       ; conditional branch to subroutine
BSCLR      ; branch to subroutine if bit clear
BSR        ; unconditional branch to subroutine
BSSET      ; branch to subroutine if bit set
DEBUG      ; enter debug mode
FBcc       ; conditional branch
FBScc      ; conditional branch to subroutine on floating-point condition
FFcc       ; conditional data ALU operation without SR update
FFcc.U     ; conditional data ALU operation with SR update
FJcc       ; conditional jump
FJScc      ; conditional jump to subroutine
FTRAPcc    ; conditional software interrupt
IFcc       ; conditional data ALU operation without SR update
IFcc.U     ; conditional data ALU operation with SR update
ILLEGAL    ; illegal instruction interrupt
Jcc        ; conditional jump
JCLR       ; jump if bit clear
JMP        ; unconditional jump
JScc       ; conditional jump to subroutine
JSCLR      ; jump to subroutine if bit clear
JSET       ; jump if bit set
JSR        ; unconditional jump to subroutine
```

```
JSSET       ; jump to subroutine if bit set
NOP         ; no operation
RESET       ; reset I/O peripheral devices
RTI         ; return from interrupt service routine
RTR         ; return from subroutine and restore SR
RTS         ; return from subroutine
STOP        ; stop processing and enter low-power stand-by mode
TRAPcc      ; conditional software interrupt
WAIT        ; wait for interrupt and enter low-power stand-by mode
```

4.4.5 Host interfacing

As introduced in subsection 4.4.1, the DSP96002 includes two ports (A and B) for external interfacing of both memory and other devices. These ports include address, data, and control buses. Each of these ports has associated with it a Host Interface (HI) which provides a 32-bit parallel port for connection of the DSP96002 to a host processor or DMA (Direct Memory Access) controller [DSP89]. These host interfaces are provided to support a wide variety of applications, including multiprocessor interfacing. Each HI includes memory-mapped I/O registers for data, control, and status information.

Support for multiprocessing is provided by a set of *host functions* on each HI. The external host processor or DMA controller communicates with the DSP96002 HI via memory-mapped calls to these functions. The host processor may be another DSP96002 or almost any other type of processor. The DSP96002 itself views each HI as a memory-mapped I/O device which occupies four 32-bit words in the X address space. This allows DSP96002 code to access the HI like any other memory-mapped I/O device using polling, interrupts, or DMA techniques.

From the many data, address, and control lines provided with each DSP96002 port, a subset of these are used in HI mode. These include the entire bidirectional data bus (D0-D31), five control lines (R/W*, TS*, HS*, HA*, and HR*), and four of the thirty-two address lines (A2-A5). The data bus is of course used for 32-bit bidirectional transfers, while the address bus subset serves as a selector for choosing one of sixteen potential host functions (some of which being reserved for future expansion). The five control lines are described below for HI operations:

R/W* This pin acts as the read/write input to each HI, and is high for a read access and low for a write access.

TS* This active-low transfer strobe pin is used as an input to enable data bus output drivers during host processor to DSP96002 read operations and to latch data in the HI during host processor to DSP96002 write operations.

HS* This active-low host select input pin is asserted low to enable selection of the HI functions by the address lines A2-A5. When TS* is asserted while HS* is asserted, a data transfer takes place through the HI.

HA* This active-low host acknowledge input pin is used to acknowledge receipt back to the host processor of either an interrupt request or a DMA request.

HR* This active-low host request output pin is used to indicate when the HI is requesting service by an external device. This pin is typically connected to an interrupt request pin on the servicing device, such as another processor or DMA controller.

In order to provide support for high-speed interprocessor communication, each HI consists of a double-buffered set of registers. Each set consists of two sides: one to be used by the DSP96002; and the other for use by the host processor via HI function calls. A block diagram describing this HI register set is shown in Figure 4.4.

There are four 32-bit registers in the HI register set which are accessible to the local DSP96002. These are the HTX/HTXC, HRX, HCR, and HSR registers. Each of these registers is mapped to a unique address at or above $FFFFFF80 (known as on-chip peripheral space) in the X address space. The HTX/HTXC (Host Transmit or Host Transmit and Clear) data register is a write-only register for DSP96002 to host processor data transfers. The HRX (Host Receive) data register is a read-only register for host processor to DSP96002 data transfers. The HCR (Host Control Register) is a read/write register which is used by the DSP96002 to control HI flags and interrupts. Finally, the HSR (Host Status Register) is a read-only status register used to indicate the current status of the HI, such as whether or not the HRX register is full and whether or not the HTX/HTXC register is empty.

On the host processor side of the HI interface, there are six 32-bit registers accessible only to the host processor via memory-mapped I/O. These are the RX, TX, SEM, IVR, CVR, and ICS registers. The RX (Receive) read-only register forms the second buffer of the double-buffer for data transfers from the DSP96002 to the host processor. Similarly, the TX (Transmit) write-only register forms the first buffer of the double-buffer for data transfers from the host processor to the DSP96002. The SEM (Semaphore) register is a read/write register of semaphores for arbitrating access to shared buses in a tightly-coupled multiprocessor system. The IVR (Interrupt Vector Register) is a read/write register containing exception vector numbers for use with MC680x0-family host processor interfaces. The CVR (Command Vector Register) is a read/write register used by the host processor to request vectored interrupt service from the DSP96002. Finally, the ICS (Interrupt Control/Status Register) is a read/write register used by the host processor to control the HI and perform status checks on the current state of the HI, such as whether or not RX is full and whether or not TX is empty.

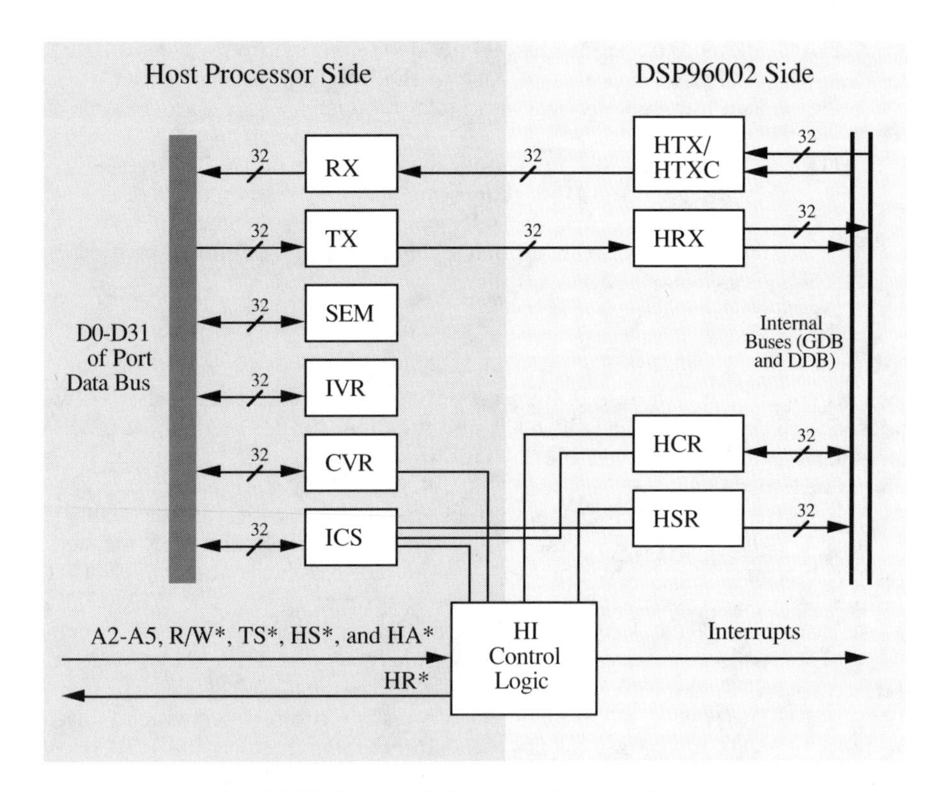

FIGURE 4.4 HI Registers for One Port

The management of the double-buffer system is provided by the HI hardware of each port. For example, when the DSP96002 sends data to the HTX/HTXC register, the HI checks the status of the RX register, and if it is empty, it moves the data from HTX/HTXC to the RX. It then changes the flag in the HSR to indicate that HTX/HTXC is now empty and also modifies the flag in the ICS to indicate that RX is now full. In this way, the DSP96002 can either poll or await an interrupt based on this HSR flag and write another value to

HTX/HTXC when it becomes free. Similarly, the host processor can either poll or await an interrupt based on this ICS flag and read the value from RX when it becomes filled.

The host processor communicates with the HI on the DSP96002 port via host function calls. These function calls are encoded on port address lines A2-A5 along with R/W* as inputs to the DSP96002. Some of these functions include: *TX register write*; *RX register read*; *ICS register read*; and *ICS register write*. For example, in order for a host processor to send information to the DSP96002, it might first begin by requesting an *ICS register read*. If the flag in the ICS indicates that the TX register is empty, the host might then proceed with a *TX register write* request. Otherwise, the host might wait until the TX register is eventually emptied by the HI hardware into the HRX register on the DSP96002 side of the interface.

4.5 Summary

This chapter has presented an overview to the field of digital signal processing and DSP implementation. DSP is concerned with the processing of a wide variety of signals using advanced digital techniques and devices. The importance of DSP continues to increase with no visible sign of saturation. DSP applications are widespread and include such areas and biomedical systems, acoustics, sonar, radar, seismology, speech synthesis and recognition, image processing, and data communication.

While the theory and algorithms for DSP are numerous, most DSP applications are related to five primary operations. Of these, the operations of digital filtering and fast Fourier transformations are particularly representative of DSP computational requirements. Digital filters are generally used to select certain frequencies of a signal to pass and certain ones to attenuate. Fast Fourier transform algorithms represent an efficient mechanism for performing discrete Fourier transforms, which are generally used for conversions between the time and frequency domains.

One of the most recent and important developments in DSP is the DSP microprocessor. These devices are capable of performing numerically-intensive DSP tasks in a highly efficient manner. DSP microprocessors employ a number of architectural strategies to achieve these performance improvements. Three of the most important of these strategies are specialized arithmetic hardware for high-speed operations, multiple buses for parallel memory access, and internal pipelining for parallel execution.

One of the most advanced of these microprocessors is the Motorola DSP96002. This microprocessor implements the IEEE 754-1985 floating-point standard in hardware, and features an extended Harvard architecture with three separate address spaces. Two complete sets of 32-bit external ports are provided, as are on-chip RAM and ROM units for embedded applications.

Versatile and sophisticated instructions, addressing modes, registers, and host interfaces are included to support a wide variety of DSP application requirements.

Now that fault-tolerant computing, parallel computing, and digital signal processing and processor issues have been introduced, we are ready to proceed with the development of the system design. These background issues will be used to discuss and develop the implementation requirements for basic DSP operations, as well as the most effective architecture to address these requirements.

Chapter 5
SYSTEM DESIGN

In this chapter, a set of premises and goals is identified for the system, and these parameters are used as the basis for the design and development of a new fault-tolerant computing system. Topics include such considerations as DSP implementations, redundancy techniques, communication architectures, clock synchronization, operating system support, interface and communication design, and system design enhancements.

5.1 Premises and Goals

The design of this new system will be based on a number of premises and goals. The first concern of the design is the choice between real-time versus non-real-time applications. In order to potentially satisfy a wide variety of DSP application requirements, the system will be based on handling problems in a real-time manner. While not all DSP applications require real-time implementation, it is considered simpler to adapt a real-time system to non-real-time problems than the reverse.

The design will incorporate advanced microprocessors along with appropriate support hardware to form the processing elements. The use of advanced microprocessors presents a number of advantages over custom logic, such as ease of design and test, flexibility, reduced development time, availability of an existing software and hardware base, maintainability, reliability, simplicity, and cost. In particular, DSP microprocessors will be used, as it is felt that in order to design a system that performs DSP operations efficiently and effectively, it makes sense to start with advanced processors that are individually designed for exactly that. Due to its unique combination of capabilities in DSP performance, dual ports, and multiprocessor simulation support, the DSP96002 has been selected as the microprocessor of choice for this system design [GEOR92].

The input and output data for this system will be based on a simple data acquisition model. For example, the digital input may be treated as immediate output from one or more A/D converters, and the digital output may be assumed to drive one or more D/A converters. In this way, input data is sampled at a rate equal to or exceeding the capabilities of the system, the data is processed by the system, and the results are written as output.

The next concern is the fault-tolerance goals of the system. When considering high-availability, long-life, and critical-computation goals (as illustrated in section 2.7), clearly critical-computation best describes the needs of most DSP problems. This in turn leads to a system which will place primary fault-tolerance emphasis on a high level of reliability. As introduced in section 4.3, the primary operations which will serve as representatives for potential DSP applications are digital filtering and fast Fourier transformations. In both cases, reliability and performance are key issues. And since some applications demand higher reliability than others, this system will be characterized as a dynamic balance between reliability and performance, with cost factors included at each step.

In order to develop a reliable system, it makes sense to start with reliable components. To better achieve this goal, off-the-shelf hardware devices will be employed in the design wherever possible, since component reliability increases with the maturity of the device. Simplicity of design will be stressed throughout the system, since simpler designs are more easily verified and understood, making potential fault-tolerance achievements more predictable. At each step, individual shared elements of the system will be minimized, in the hope that this will lead to minimal single points of failure.

Due to the real-time constraints imposed by such operations as digital filtering, the system must be able to tolerate a permanent or transient fault in a manner which avoids any degradation in performance. That is, virtually no single permanent or transient hardware fault should be capable of impeding the operation of the system.

In order to understand the fault-tolerance goals of this system, the sources of failures must first be considered. Failures in computing systems occur due to a number of different sources. Primary sources can be divided into the following five categories: processor and memory failure; communication network failure; peripheral device failure; environmental and power failure; and human error [DHIL87].

Processor errors are less frequent but potentially more catastrophic than memory errors. Failures in the communication network can often render a distributed system completely inoperable. The failure of peripheral devices, such as sensors, actuators, A/D and D/A converters, etc., while not typically catastrophic to system operation, can in fact degrade or destroy the very I/O data a real-time DSP system is intended to operate upon.

Environmental failures include problems with temperature, radio-frequency interference (RFI), electromagnetic interference (EMI), natural disasters, etc. Human errors consist primarily of software errors due to programming design and coding mistakes. While modern software engineering practices tend to improve the quality of the software and thus reduce the number of errors, the state of the art in program verification is not nearly advanced enough to eliminate all errors in production-level code.

A number of assumptions have be made in order to develop a realistic and practical fault-tolerant real-time DSP system. Perhaps most importantly, processor faults, memory faults, and other component and communication path faults are the primary emphasis for this fault-tolerant design. In failing, processors and other components may fail completely or simply emit erroneous information (i.e. become an insane node).

Since research with such systems as the FTMP and SIFT systems has resulted in documented methods for fault-tolerant power supplies, loss of power in this system will not be considered as a potential problem [SMIT86a]. Human errors will be addressed when software redundancy is considered for this system.

Many issues must be considered in order to achieve the goals described, in terms of factors such as the processors, the interconnection network, the operating system and application software, and the overall architecture. For example, what form or forms of redundancy can best be employed? Should the processors be tightly or loosely synchronized? Should the architecture be reconfigurable to either handle different classes of problems or different levels of reliability required? These and many other issues will be addressed.

5.2 DSP Implementation Considerations

In order to best determine an overall architecture for the system, we must consider potential architectures for DSP applications. There are many methods for implementing both digital filters and fast Fourier transforms. For the purpose of better understanding potential implementations in a fault-tolerant manner, we begin by considering conventional approaches to our two basic DSP operations.

5.2.1 Digital filtering

By using algebraic manipulation, the form of the defining difference equations (DEs) of a digital filter can be altered so as to modify the overall requirements of its implementation [GEOR88]. The following four LTI IIR digital filter difference equation representations as presented by Phillips and Nagle [PHIL84] are examined (where order n is the maximum of the numerator and denominator orders of the transfer function, $x(k)$ is the current input, $y(k)$ is the current output, and $m(k)$, $p_i(k)$, $r_i(k)$, and $q_i(k)$ are temporary intermediate variables used in factoring the equations). In each case, the initial conditions of the difference equations are determined by the application requirements (e.g. all set to zero):

1D realization (also known as Direct Form II in [BOWE82]):

$$m(k) = x(k) - b_1 m(k-1) - b_2 m(k-2) - \dots - b_n m(k-n) = x(k) - \sum_{i=1}^{n} b_i m(k-i) \quad (5.1)$$

$$y(k) = a_0 m(k) + a_1 m(k-1) + a_2 m(k-2) + \dots + a_n m(k-n) = \sum_{i=0}^{n} a_i m(k-i) \quad (5.2)$$

2D realization (the transpose of 1D):

$$p_i(k) = p_{i+1}(k-1) + a_i x(k) - b_i y(k) \text{ , for } i = 1 \text{ to } n-1 \quad (5.3)$$

$$p_n(k) = a_n x(k) - b_n y(k) \quad (5.4)$$

$$y(k) = a_0 x(k) + p_1(k-1) \quad (5.5)$$

3D realization (also known as Direct Form I in [BOWE82]):

$$y(k) = a_0 x(k) + a_1 x(k-1) + \dots + a_n x(k-n) - b_1 y(k-1) - b_2 y(k-2) - \dots - b_n y(k-n)$$

$$\text{or equivalently} \quad y(k) = \sum_{i=0}^{n} a_i x(k-i) - \sum_{i=1}^{n} b_i y(k-i) \quad (5.6)$$

4D realization (the transpose of 3D):

$$r_0(k) = x(k) + r_1(k-1) \quad (5.7)$$

$$q_n(k) = a_n r_0(k) \quad (5.8)$$

$$r_n(k) = -b_n r_0(k) \quad (5.9)$$

$$q_i(k) = a_i r_0(k) + q_{i+1}(k-1) \text{ , for } i = 1 \text{ to } n-1 \quad (5.10)$$

$$r_i(k) = -b_i r_0(k) + r_{i+1}(k-1) \text{ , for } i = 1 \text{ to } n-1 \quad (5.11)$$

$$y(k) = a_0 r_0(k) + q_1(k-1) \quad (5.12)$$

In addition, the following is the direct form for an LTI FIR digital filter:

$$y(k) = a_0 x(k) + a_1 x(k-1) + \ldots + a_n x(k-n) = \sum_{i=0}^{n} a_i x(k-i) \tag{5.13}$$

Uniprocessor Implementation

With a typical uniprocessor implementation (without internal parallelism) of the four IIR digital filter DEs presented, all of the floating-point arithmetic calculations are performed in a sequential manner, one after the other. As can be seen from the realization equations previously shown, the operations required primarily consist of multiplications, additions, and the loading and storing of coefficients and previous inputs and outputs. By counting up the total number of each of these types of operations, the information in Table 5.1 has been generated.

TABLE 5.1 Approximate Uniprocessor Operation Counts for Order *n* LTI IIR Digital Filters

FORM	2-Input Multiplies	2-Input Additions	Delay (memory) Elements	Memory/Register Accesses (excluding intermediate access)
1D	$2n+1$	$2n$	n	$4n+5$
2D	$2n+1$	$2n$	n	$6n+3$
3D	$2n+1$	$2n$	$2n$	$4n+3$
4D	$2n+1$	$2n$	$2n$	$8n+5$
FIR	$n+1$	n	n	$2n+3$

Therefore the overall timing approximation equations of the DE forms considered are as follows:

$$t_{1D} = (2n+1)t_{mult} + 2nt_{add} + (4n+5)t_{mem} \tag{5.14}$$

$$t_{2D} = (2n+1)t_{mult} + 2nt_{add} + (6n+3)t_{mem} \tag{5.15}$$

$$t_{3D} = (2n+1)t_{mult} + 2nt_{add} + (4n+3)t_{mem} \tag{5.16}$$

$$t_{4D} = (2n+1)t_{mult} + 2nt_{add} + (8n+5)t_{mem} \tag{5.17}$$

$$t_{FIR} = (n+1)t_{mult} + nt_{add} + (2n+3)t_{mem} \tag{5.18}$$

where:

t_{mult} = total time to multiply two floating-point numbers

t_{add} = total time to add two floating-point numbers

t_{mem} = total time to access a floating-point no. via memory/register

The implementation of any of these DE forms can be readily done via the use of one of the many general-purpose microprocessors available, such as one from the Intel 80x86 or the Motorola 680x0 families of microprocessors. For example, consider using a hypothetical microprocessor with the following execution times to implement a filter of order eight (i.e. $n = 8$) :

$$t_{mult} = 10\ \mu\text{sec},\ t_{add} = 4\ \mu\text{sec},\ t_{mem} = 1\ \mu\text{sec}$$

By substituting into the overall general timing approximation equations:

t_{1D} = 271 μsec => 3.69 kHz maximum sample frequency

t_{2D} = 285 μsec => 3.51 kHz maximum sample frequency

t_{3D} = 269 μsec => 3.72 kHz maximum sample frequency

t_{4D} = 302 μsec => 3.30 kHz maximum sample frequency

t_{FIR} = 141 μsec => 7.09 kHz maximum sample frequency

Thus, for a sample to signal frequency ratio of two, the best IIR general-purpose microprocessor implementation in this example (3D) can be used for signals of at most 1.86 kHz. While the FIR information of course looks more capable, it should be noted that this is not a fair comparison, since for an FIR filter to match the features of an IIR filter, it normally requires a much higher order. Thus, the FIR data is included at this point more as a means of clarification than comparison.

Due to the increased coefficient sensitivity of the digital filter difference equations presented as the order n increases, the overall defining equation(s) are often divided into separate second-order sections either by factoring the transfer functions associated with the difference equations or by applying partial-fraction expansion to them. Examples of both factored and expanded second-order sections are shown in Figures 5.1 and 5.2 respectively (where P is the ceiling of $n/2$).

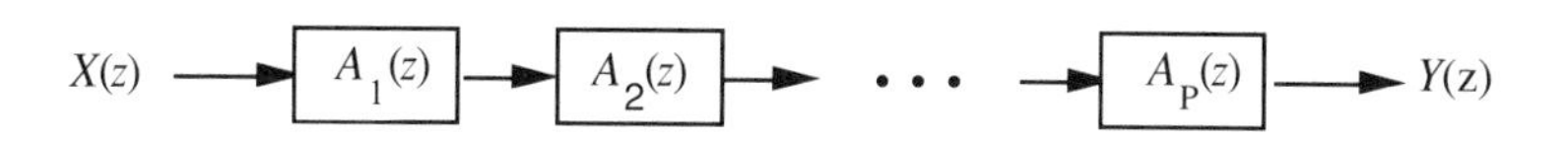

FIGURE 5.1 Factored Second-Order Sections

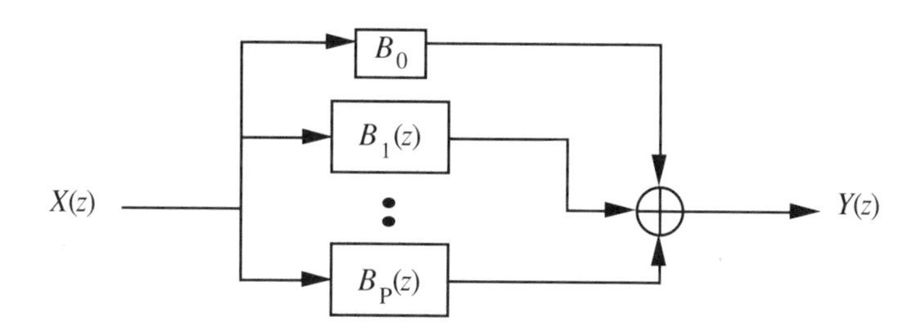

FIGURE 5.2 Expanded Second-Order Sections

In order to address digital filtering requirements which exceed the capabilities of uniprocessor implementations, these two methods for separating the defining equations of the filters into sections can be used to form the basis with which parallel implementation can be achieved. By factoring the defining equations, a parallel computer system consisting of processors organized into an MIMD linear array can be used to match the topology of the algorithm. Similarly, by using partial-fraction expansion, a parallel system of processors organized into an SIMD array processor will fit the algorithm structure.

MIMD Linear Array Implementation

The overall general timing approximation equations developed previously can be extended to the case whereby factored second-order sections are used to implement the digital filter on a single processor. This is done by using the appropriate equations (for 1D, 2D, etc.) where $n = 2$, multiplying the result by the number of second-order sections required, and accounting for the fact that the output of each section in turn becomes the input to the next section by including a memory access between sections. Consider the following equations for a digital filter of order eight:

$$t_{1D} = \left\{ \left[(2n+1)t_{mult} + 2nt_{add} + (4n+5)t_{mem}\right]_{n=2} \right\} * 4 + 3t_{mem}$$

$$= \{ 5t_{mult} + 4t_{add} + 13t_{mem}\} * 4 + 3t_{mem}$$

$$= 20t_{mult} + 16t_{add} + 55t_{mem} \quad (5.19)$$

$$t_{2D} = 20t_{mult} + 16t_{add} + 63t_{mem} \quad (5.20)$$

$$t_{3D} = 20t_{mult} + 16t_{add} + 47t_{mem} \quad (5.21)$$

$$t_{4D} = 20t_{mult} + 16t_{add} + 87t_{mem} \quad (5.22)$$

$$t_{FIR} = 12t_{mult} + 8t_{add} + 31t_{mem} \quad (5.23)$$

By implementing a filter of order eight using the hypothetical microprocessor discussed previously, the following is obtained:

t_{1D} = 319 μsec => 3.14 kHz maximum sample frequency

t_{2D} = 327 μsec => 3.06 kHz maximum sample frequency

t_{3D} = 311 μsec => 3.22 kHz maximum sample frequency

t_{4D} = 351 μsec => 2.85 kHz maximum sample frequency

t_{FIR} = 183 μsec => 5.47 kHz maximum sample frequency

Thus, in this case, for the same sample to signal frequency ratio as before, the best IIR digital filter implementation in this example (3D) can be used for signals of at most 1.61 kHz.

As presented in [PHIL84], an 8086 running at a clock frequency of 5 MHz that is used to implement an IIR digital filter of order eight using second-order factored sections (and running the code developed by Phillips and Nagle) indicates the following timing constraints:

t_{1D} = 791 μsec => 1.26 kHz maximum sample frequency

t_{2D} = 785 μsec => 1.27 kHz maximum sample frequency

$t_{3D} = 757\ \mu\text{sec}$ => 1.32 kHz maximum sample frequency

$t_{4D} = 777\ \mu\text{sec}$ => 1.28 kHz maximum sample frequency

Thus, for the same sample to signal frequency ratio as before, a 5 MHz 8086 implementing a IIR digital filter of order eight (in 3D form) can handle at most signals of approximately 0.66 kHz.

In order to satisfy digital filtering performance requirements which may exceed the capabilities of *any* uniprocessor implementation, it becomes necessary to incorporate some form of parallel processing and parallel architecture. One such architecture which can be used to introduce parallelism into the execution of a DSP process is the MIMD linear processor array.

The MIMD linear array architecture is particularly applicable to the parallel implementation of digital filters, since its topology matches that of the factored second-order section realizations. Like the instruction pipelines used in most modern microprocessors, linear processor arrays tend to subdivide a task into a sequence of subtasks [HWAN84]. However, in an MIMD linear array, each of these subtasks operates concurrently with and independent of the other subtask stages in powerful and independent processors. An illustration of how a task is separated into multiple linear array subtasks is shown in Figure 5.3.

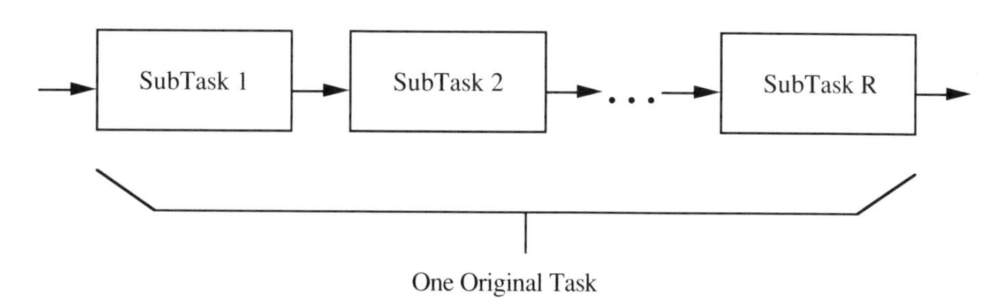

FIGURE 5.3 Separating a Task into R Subtasks

Of course, for linear arrays to be effective, it must be possible and practical to break the necessary task into subtasks. While many real world applications cannot always meet this criteria, the implementation of factored second-order sections of digital filters lends itself to this type of architecture.

While the overall latency of the system is increased, throughput is comparable to that of an individual processor stage once the array has filled. By implementing m second-order sections on each of p processors in the array, and including incoming and outgoing data interprocessor data transfer, the following overall timing approximation equations for the entire system can be used once the array has filled (where p is the ceiling of $n/(2m)$):

$$t_{1D} = \left\{ \left[(2n+1)t_{mult} + 2nt_{add} + (4n+5)t_{mem} \right]_{n=2} \right\} * m + (m-1)t_{mem} + 2t_{comm}$$

$$= \{ 5t_{mult} + 4t_{add} + 13t_{mem} \} * m + (m-1)t_{mem} + 2t_{comm}$$

$$= 5mt_{mult} + 4mt_{add} + (14m - 1)t_{mem} + 2t_{comm} \quad (5.24)$$

$$t_{2D} = 5mt_{mult} + 4mt_{add} + (16m - 1)t_{mem} + 2t_{comm} \quad (5.25)$$

$$t_{3D} = 5mt_{mult} + 4mt_{add} + (12m - 1)t_{mem} + 2t_{comm} \quad (5.26)$$

$$t_{4D} = 5mt_{mult} + 4mt_{add} + (22m - 1)t_{mem} + 2t_{comm} \quad (5.27)$$

$$t_{FIR} = 3mt_{mult} + 2mt_{add} + (8m - 1)t_{mem} + 2t_{comm} \quad (5.28)$$

where

t_{comm} = total time to perform an interprocessor floating-point transfer

For example, consider a transfer function for an IIR digital filter of order eight which has been broken down into second-order sections via factoring:

$$\frac{Y(z)}{X(z)} = D_1(z)\, D_2(z)\, D_3(z)\, D_4(z) \quad (5.29)$$

where each $D_i(z)$ is of the form:

$$D_i(z) = \frac{\alpha_{0i} + \alpha_{1i}z^{-1} + \alpha_{2i}z^{-2}}{1 + \beta_{1i}z^{-1} + \beta_{2i}z^{-2}} \quad (5.30)$$

In this case, a four-stage linear array of processors can be used for implementation, as shown in Figure 5.4. Connections between processors may be unidirectional or bidirectional as algorithm and application requirements dictate. In this case, unidirectional connections are portrayed.

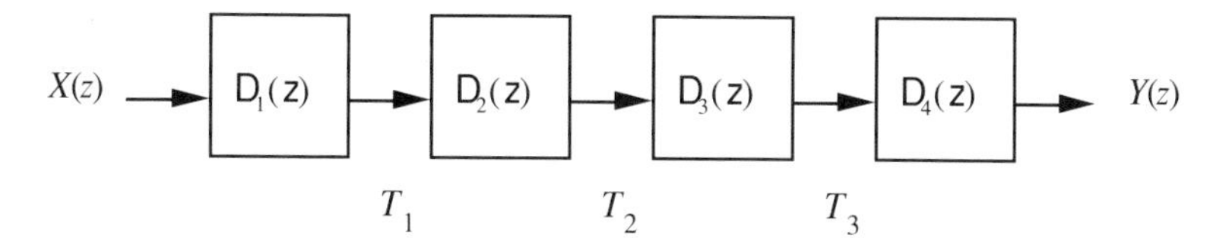

FIGURE 5.4 Example Four-Stage Linear Processor Array

This in turn implies $T_1(z) = D_1(z)\,X(z), \ldots, Y(z) = D_4(z)\,T_3(z)$

$$\Rightarrow t_1(k) = \alpha_{01}\,x(k) + \alpha_{11}\,x(k\text{-}1) + \alpha_{21}\,x(k\text{-}2) - \beta_{11}t_1(k\text{-}1) - \beta_{21}t_1(k\text{-}2) \quad (5.31)$$

$$\Rightarrow t_2(k) = \alpha_{02}t_1(k) + \alpha_{12}t_1(k\text{-}1) + \alpha_{22}t_1(k\text{-}2) - \beta_{12}t_2(k\text{-}1) - \beta_{22}t_2(k\text{-}2) \quad (5.32)$$

$$\Rightarrow t_3(k) = \alpha_{03}t_2(k) + \alpha_{13}t_2(k\text{-}1) + \alpha_{23}t_2(k\text{-}2) - \beta_{13}t_3(k\text{-}1) - \beta_{23}t_3(k\text{-}2) \quad (5.33)$$

$$\Rightarrow y(k) = \alpha_{04}t_3(k) + \alpha_{14}t_3(k\text{-}1) + \alpha_{24}t_3(k\text{-}2) - \beta_{14}\,y(k\text{-}1) - \beta_{24}y(k\text{-}2) \quad (5.34)$$

Thus, the processor used in each stage would hold its particular α and β coefficients as well as its last two previous outputs and inputs, and could then calculate the correct output value based on these and the data fed to it by the previous stage.

Since in this case the number of sections $m = 1$, the order $n = 8$, and the number of processors $p = 4$, and using the hypothetical microprocessors discussed previously where $t_{\text{comm}} = 2$ µsec, the following is obtained:

t_{1D} = 83 µsec => 12.05 kHz maximum sample frequency

t_{2D} = 85 µsec => 11.77 kHz maximum sample frequency

t_{3D} = 81 µsec => 12.35 kHz maximum sample frequency

t_{4D} = 91 µsec => 10.99 kHz maximum sample frequency

t_{FIR} = 49 µsec => 20.41 kHz maximum sample frequency

Thus, in this case, for the same sample to signal frequency ratio as before, the best four-processor IIR digital filter linear array implementation in this example (3D) can be used for signals of at most 6.18 kHz.

An MIMD linear array approach to the implementation of digital filters is one that could potentially be used on a wide variety of processors, depending upon the interconnections supported. For example, transputers have been shown to be a viable candidate for digital filtering with linear processor arrays [GEOR89a]. In particular, the DSP96002 is an especially promising microprocessor platform for this architecture due to its dual port organization, simulation support, and DSP support [GEOR92].

As introduced in section 3.2, two types of MIMD linear processor arrays are the systolic and wavefront architectures. Although these architectures are normally associated with VLSI computing structures with simple processor cells (i.e. chip-level multiprocessors), their techniques can also be applied to a linear array of sophisticated and powerful microprocessors like the DSP96002. A *systolic* array consists of a network of processors which routinely and rhythmically compute and pass data through the system using global clock synchronization [KUNG82a]. A *wavefront* array features the same regularity, modularity, and local communication properties as systolic arrays, but operates in a more asynchronous manner. Information is passed between processors in a data-driven rather than clock-driven manner [KUNG82b].

SIMD Array Processing Implementation

Another architecture which can be used to introduce parallelism into the execution of a DSP process is the SIMD array processor. Like the MIMD linear array, this architecture is also particularly applicable to the parallel implementation of digital filters, since its topology matches that of the expanded second-order section realizations.

An SIMD array processor is a synchronous array of parallel processors all controlled by a central master control unit (CU). An array processor is one which operates in a single-instruction multiple-data (SIMD) fashion. In this way, at any given time, every processing element (PE) in an array processor is executing the same instruction, only on different data [HWAN84].

For array processing techniques to be effective, it must be possible and practical to break the necessary task into disjoint identical operations that can be fit to an SIMD array processing architecture. While many real world applications cannot always meet this criteria, the implementation of expanded second-order sections for digital filters lends itself to this type of architecture. This is accomplished by operating upon the transfer function using partial-fraction expansion [OPPE75].

For example, consider a transfer function for an LTI IIR digital filter of order eight which has been broken down into second-order sections via partial-fraction expansion:

$$\frac{Y(z)}{X(z)} = D_0 + D_1(z) + D_2(z) + D_3(z) + D_4(z) \tag{5.35}$$

where D_0 is a constant (zero if the order of the transfer function denominator exceeds that of the numerator), and each $D_i(z)$ is of the form:

$$D_i(z) = \frac{\alpha_{1i}z^{-1} + \alpha_{2i}z^{-2}}{1 + \beta_{1i}z^{-1} + \beta_{2i}z^{-2}} \tag{5.36}$$

In this case, a four-PE array processor (where each PE and the CU are microprocessors) could be used for implementation, such as the array processor shown in Figure 5.5.

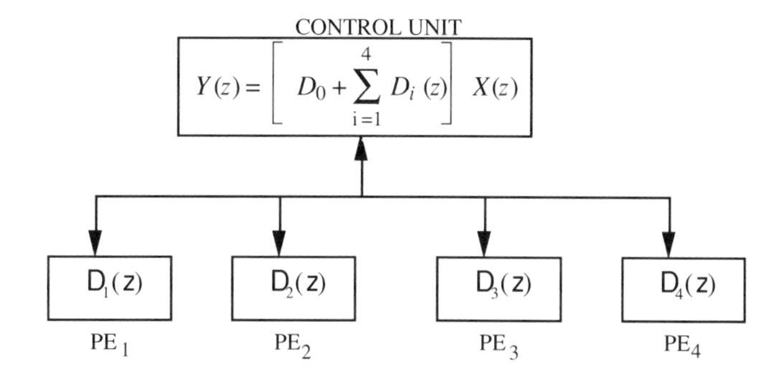

FIGURE 5.5 Example Array Processor with Four Processing Elements

Then each

$$D_i(z) \Rightarrow y_i(k) = \alpha_{1i}x(k-1) + \alpha_{2i}x(k-2) - \beta_{1i}y_i(k-1) - \beta_{2i}y_i(k-2) \tag{5.37}$$

and

$$y(k) = D_0 + \sum_{i=1}^{4} y_i(k) \tag{5.38}$$

Thus, each PE would hold its particular α and β coefficients as well as its last two previous outputs and inputs, and could calculate the correct output value based on these and the current input fed to it by the CU. The CU in turn would collect all of the $y_i(k)$ terms of the $y(k)$ output, and sum them up along with the D_0 term.

In terms of processor interaction dependencies, assuming all of the PEs possess their coefficients, the process begins with the CU providing the input to all PEs. Then all of the PEs operate in parallel, computing their outputs. These outputs in turn are sent back to the CU for subsequent summation. Now, by counting the operations required, a general timing approximation equation can be generated where each PE implements m cascaded second-order sections (where as before p is the ceiling of $n/(2m)$):

$$t_{AP} = 4mt_{mult} + (p + 3m)t_{add} + (p + 9m + 3)t_{mem} + (p + 1)t_{comm} \tag{5.39}$$

If we implement the last example using our hypothetical microprocessors, then $n = 8$, $m = 1$, $p = 4$, and:

$$t_{AP} = 4(1)t_{mult} + (4 + 3(1))t_{add} + (4 + 9(1)+3)t_{mem} + (4 + 1)t_{comm}$$

$$= 94\ \mu\text{sec} \implies 10.64\ \text{kHz maximum sample frequency}$$

Thus, in this case, for the same sample to signal frequency ratio as before, the four-PE IIR digital filter array processor implementation in this example can be used for signals of at most 5.32 kHz. As this example illustrates, when compared to the MIMD linear array implementation, array processing requires more overhead in terms of communication delay between the CU and the PEs. Of course, some optimizations are possible to narrow this gap, such as having one or more PEs perform some or all of the $y_i(k)$ summations to determine the overall output $y(k)$.

Digital Filtering Implementation

Clearly, the means by which a digital filter may be implemented are extensive. A single general-purpose processor may be used, with such advantages as easier programming and interfacing. A single special-purpose processor may be used for faster sampling rates. And, by combining processors together in various ways based on the modern techniques in computer architecture, even greater sampling rates may be achieved by introducing and increasing the amount of parallelism in the execution. In the parallel computing architectures of today, it is clear that the primary problem at this point is mapping the software and applications onto the hardware. However, with an application such as a digital filter implementation, where the data dependencies and the execution needs can be determined in advance, the full parallelism of the task can be achieved.

Based on the overhead involved, the parallel implementation of digital filters via MIMD linear arrays for factored second-order sections appears to be the most favorable approach. With SIMD array processing for expanded second-order sections, information must be passed between the CU and the PEs routinely. Since we have shown that digital filters can be partitioned in such a way as to imply minimal overhead and maximal uniformity, linear arrays appear to be the best selection with respect to simplicity and performance for DSP operations like digital filters. This may be in the form of single or multiple linear arrays using either the systolic or wavefront array model. The performance issues of DSP operations like fast Fourier transforms remain to be considered, as do implicit and explicit fault-tolerance issues.

5.2.2 Fast Fourier transformations

One common approach to implementing discrete Fourier transforms using the fast Fourier transform (FFT) algorithm, which was introduced earlier, is known as the *decimation-in-time* technique, due to the way the time domain input sequence is divided or *decimated* into successively smaller sequences. When used in base two, this complex N-point fast Fourier transformation algorithm requires $(N/2)\log_2 N$ butterfly operations. Each butterfly operation has two inputs and two outputs u and v, and is based on the following two equations:

$$u(k+1) = u(k) + W_N^r v(k) \tag{5.40}$$

$$v(k+1) = u(k) - W_N^r v(k) \tag{5.41}$$

The variable W_N is known as the *twiddle factor,* where:

$$W_N^r = e^{-j\frac{2\pi}{N}r} = \cos\frac{2\pi}{N}r - j\sin\frac{2\pi}{N}r \tag{5.42}$$

Each twiddle factor term is a complex coefficient which can be referenced from trigonometric lookup tables at execution time by using positive integer r as an index (where the value of r depends upon which stage or pass of the FFT the butterfly belongs to [BRIG74]) . In terms of arithmetic requirements, the butterfly consists of one complex multiplication and two complex additions, which in turn dictates a total of four real multiplications and six real additions. By including memory or register variable access in a similar manner to that previously done with digital filter difference equations, the following butterfly timing approximation equation can be generated:

$$t_{\text{bfly}} = 4t_{\text{mult}} + 6t_{\text{add}} + 8t_{\text{mem}} \tag{5.43}$$

Uniprocessor Implementation

With a typical uniprocessor implementation (without internal parallelism) of the fast Fourier transformation presented, all of the butterfly calculations are performed in a sequential manner, one after the other. Thus:

$$t_{SP} = (N/2)(\log_2 N)\, t_{bfly}$$

$$= (N/2)(\log_2 N)(4t_{mult} + 6t_{add} + 8t_{mem}) \tag{5.44}$$

For example, consider implementing a 1024-point FFT using the hypothetical microprocessor previously used, where as before:

$$t_{mult} = 10\ \mu sec,\ t_{add} = 4\ \mu sec,\ t_{mem} = 1\ \mu sec,\ t_{comm} = 2\ \mu sec$$

By substituting into the timing approximation equation with $N = 1024$:

$$t_{SP} = (1024/2)(\log_2 1024)(4(10) + 6(4) + 8(1)) = 512\,(10)\,(72)$$

$$= 368.64 \text{ msec}$$

MIMD Linear Array Implementation

Since the decimation-in-time base two FFT consists of $\log_2 N$ stages, where each stage is responsible for $N/2$ butterfly operations, an MIMD linear processor array implementation of the FFT is a natural approach. In particular, by matching one processor to each stage and having each processor iteratively execute the $N/2$ butterfly operations of it, the $\log_2 N$ processors needed have the potential to provide a performance improvement approaching that of a linear increase once the array has filled.

For example, consider a 1024-point FFT. If we match one processor per stage, this implies ten processors, where each processor iteratively executes 512 butterfly operations, as shown in Figure 5.6.

While the separation of tasks avoids shared memory, and thus helps to avoid a single point of failure, the amount of data to be passed between stages is proportional to the size of the FFT desired. In this case, 1024 data values must be sent and received between each stage, which implies the following overall timing approximation equation once the array has filled:

$$t_{LA} = (N/2)t_{bfly} + 2(N)t_{comm} \tag{5.45}$$

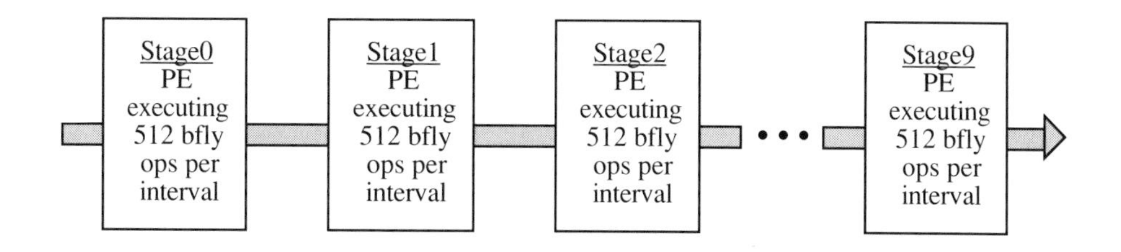

FIGURE 5.6 A 1024-point FFT using a Linear Processor Array

Using the hypothetical microprocessor timing specifications, this leads to:

$$t_{LA} = (1024/2)t_{bfly} + 2(1024)t_{comm}$$

$$= 512(72) + 2048(2)$$

$$= 40.96 \text{ msec}$$

Compared to the uniprocessor result of 368.64 msec required, using a linear array of ten processors to implement this 1024-point FFT may potentially provide an almost ten-fold improvement. Since the butterfly operation times may have more significance in the overall timing than the data transfer times as the FFT size grows, the overhead incurred may decrease as the size increases. Of course, techniques can be applied to help minimize the overhead associated with the data transfers between array stages, such as input and output buffers operating in parallel.

In a fashion similar to that carried out with digital filtering, SIMD array processing can be shown to imply increased levels of complexity and overhead. In fact, as noted in [GILB85], SIMD configurations rarely yield performance enhancements comparable to alternate approaches in real-time DSP applications.

These alternate approaches, as previously discussed, include MIMD linear processor arrays based on either the systolic or wavefront models. These models appear to be the most promising architectures available for a wide variety of real-time DSP applications, in both linear and hexagonal configurations [WHIT85]. However, it remains to be seen how these architectures can fit into the overall fault tolerance and reliability requirements desired for this system.

5.3 Redundancy Considerations

As described in chapter 2, redundancy can be incorporated using a variety of methods. These methods are based on the categories of hardware, time, software, and information redundancy.

When considering hardware and time redundancy, Kopetz has described the capability of these schemes in a real-time system [KOPE85]. In particular, as shown in Table 5.2, static hardware redundancy looks the most promising for the requirements of real-time DSP computing systems, with time redundancy limited to situations when reserve processor resources are available.

TABLE 5.2 Hardware and Time Redundancy Effects

Redundancy Type	Errors Handled			Effect on Real-Time DSP Performance
	Permanent	Transient	Intermittent	
Static Hardware Redundancy	√	√	partially	minimal
Dynamic Hardware Redundancy	√	√	partially	significant
Time Redundancy	no	√	√	significant

While dynamic hardware redundancy does indeed provide the necessary support for both permanent and transient faults, its effect on high-speed real-time performance in general, and DSP real-time performance in particular, is unsatisfactory. Unlike the fault masking provided by static redundancy techniques, dynamic redundancy attempts to detect and remove faulty hardware from operation. It is this overhead, in terms of fault detection, location, and recovery, that can interrupt and adversely effect time-critical DSP processing.

While the general form of static hardware redundancy is the NMR approach, the TMR subset of this approach is most appropriate for this system. The use of triplicated logic in TMR facilitates the tolerance of single permanent or transient faults, which directly addresses the fault-tolerance goals of this system as described in section 5.1.

Like time redundancy, the software redundancy techniques of consistency and capability checks will be limited to those situations when idle processor time is available. However, software replication will for the most part be excluded, as this technique is by far the most expensive to employ and is only cost-effective for the most critical of all applications. Instead, as done with many fault-tolerant machines, potential errors in software design will be minimized by fault avoidance using careful design, testing, and the application of software engineering principles as needed. Since the software routines for this fault-tolerant real-time DSP system may not be nearly as lengthy as those of other systems, this appears to be the best tradeoff at this point. In addition, the use of compact and carefully designed software routines is generally more amenable to formal proof techniques.

Finally, since the use of off-the-shelf devices is a fundamental requirement for this system, information redundancy employment will be limited in application so as not to impede upon real-time performance. Like time redundancy, it will be restricted to situations where reserve processing potential is available, since techniques like error-correcting codes can often imply significant additional delays and the avoidance of shared-memory in the design considerations thus far reduces the advantages of this technique.

Closely related to information and hardware redundancy is the technique known as algorithm-based fault tolerance. As described in subsection 2.3.3, algorithm-based fault tolerance methods attempt to provide fault detection and correction mechanisms specific to a particular algorithm (e.g. matrix multiplication). While this technique represents a promising approach for application-specific fine-grained parallel processing VLSI circuits, this technique currently does not provide the fault masking, flexibility, versatility, and real-time performance support desired for this generalized microprocessor-based DSP system.

5.4 Communication Architecture Considerations

Based on the premises and goals in section 5.1 and the developments in sections 5.2 and 5.3, the most promising system configuration is the linear array DSP96002-based multiprocessor architecture featuring TMR static hardware redundancy for fault masking. This architecture is proposed for use with high-speed real-time DSP applications like digital filtering and fast Fourier transforms in a fault-tolerant manner.

The system itself can be considered a distributed one, since the linear array multiprocessor configuration proposed is based on each processor possessing its own local memory with no shared-memory being used other than perhaps a buffer between stages. When considering the performance issues associated with a real-time DSP system, clearly the system needs to be fault-tolerant in a manner such that no appreciable time is spent in reconfiguration. In fact, it was this requirement that led to the choice of static hardware redundancy to begin with. And, since the DSP96002 provides as its only high-speed input/output mechanism a pair of 32-bit ports, the number and method of interconnections is limited.

As overviewed in chapter 2, there are seven basic categories of fault-tolerant multiprocessor communication architectures. These include reliable shared buses, shared-memory interconnection networks, loop architectures, tree networks, dynamically reconfigurable networks, binary cube interconnection networks, and graph networks. Each of these categories can be considered with respect to the single-fault reliability-oriented fault-tolerance requirements of this system, the real-time performance requirements of this system, and the capabilities provided by the processors to be used as the basis of this system (i.e. the Motorola DSP96002).

With reliable shared buses, a single bus is shared by all the processors via time-division multiplexing. However, this model is unsatisfactory in meeting the desired specifications for this system, in terms of both performance and fault tolerance. The use of a single bus for all interprocessor communication would serve as both a bottleneck for data transmission as well as a critical single point of failure.

Shared-memory interconnection networks are a common technique used in multiprocessors to share part or all of available memory between all of the processors. This model is incompatible with the configuration proposed, since a more loosely-coupled architecture is preferred without shared memory. In addition, the use of these networks would also impose a performance degradation when a failure is detected, and would thus interfere with time-critical DSP operations.

The loop or ring architecture consists of processors which communicate via a circular loop in a clockwise manner, a counter-clockwise manner, or both. One of the main advantages of these architectures is their ability to withstand a faulty processor or link between processors. While this model is compatible with a linear array of processors, the method by which faults are managed is nevertheless unsatisfactory. When a processor or link fails, information is routed around the faulty component. However, the overhead involved when a failure does occur, along with the degradation in performance subsequent to a processor failure, would adversely affect real-time DSP operation.

Unlike the previous horizontal approaches, tree networks use a hierarchical organization with processors divided into different levels. This model is incompatible with the linear array orientation proposed for this system. In addition, performance degradation in the presence of a failure is also a significant concern when using this technique for real-time DSP.

Dynamically reconfigurable networks are characterized by their ability to change from a horizontal to a hierarchical configuration or vice-versa while the system is operational. Like the loop architecture, this model is also compatible with a linear array of processors but limited with respect to performance degradation and real-time DSP operations.

The binary cube interconnection network has received much attention in recent years. In this model, two processors are connected and adjacent if and only if their binary addresses differ by exactly one bit. However, as with tree networks, this model is also incompatible with the proposed linear array multiprocessor organization. In addition, the overhead involved with reconfiguration and routing in the presence of a processor failure would once again deter from high-speed real-time DSP response.

Finally, fault-tolerant graph networks are a category of communication architectures based on certain types of mathematical graph structures. These networks combine some of the more attractive features from the other models and provide a correlation between bus architectures and link architectures. However, much like dynamically reconfigurable networks, this model is also compatible with a linear array of processors but limited with respect to performance degradation and real-time DSP operations. For example, when a processor is removed from operation due to a failure, the routing distance between other pairs of processors can increase, resulting in reduced performance for time-critical operations.

It appears that none of these seven fault-tolerant multiprocessor communication architecture models can accurately describe the proposed system configuration. Each of these models is more appropriate for dynamic rather than static hardware redundancy, or is otherwise incompatible with a linear array real-time system of distributed TMR processors. Instead of masking faults, these models are more oriented toward fault detection, location, and recovery techniques to reconfigure the system for further operation after a failure. It is these techniques which impose reconfiguration and other delays, as well as graceful degradation in performance, that can be detrimental to time-critical DSP operations.

Thus, the proposed system configuration involves a communication architecture which is distinct from these standard models. The method with which TMR static hardware redundancy will be incorporated must still be considered.

5.5 Static Redundancy Considerations

As discussed in chapter 2, static hardware redundancy techniques rely on voting to provide fault masking, so that the system inherently tolerates faults. While TMR static redundancy shows the most potential for fault-tolerant real-time DSP implementations, there are still many design choices related to TMR static redundancy that must be considered for this linear array multiprocessor system. These include the consideration of faults in the voter, the location and frequency of the voting process, and the technique by which voting will be accomplished [JOHN89a].

With any fault-tolerant system, a potential single point of failure is of concern. For systems based on TMR, the voter itself represents such a point.

If the voter fails, the system itself fails, despite the redundancy in processing elements that is provided (see Figure 5.7).

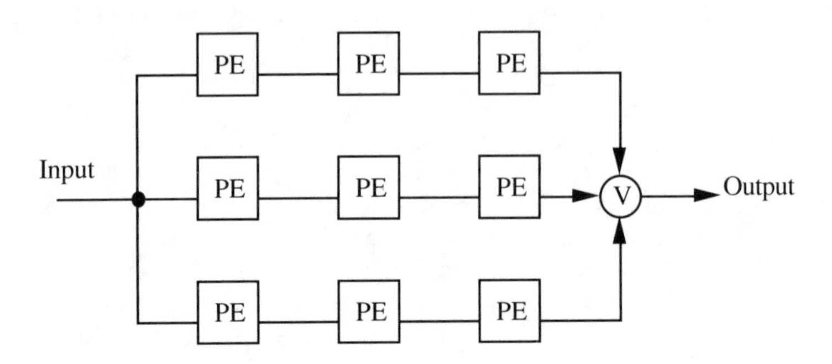

FIGURE 5.7 A TMR Linear Array with a Single Voter

One technique that has been described to eliminate this single point of failure is the use of triplicated voters. In this way, TMR is used not only for the modules in the system, but for the voter itself. This is illustrated for a simple TMR linear array in Figure 5.8. However, since most applications eventually require a single output result, this approach requires some form of single-output fault-tolerant selector or voter to either replace or augment the triplicated voters themselves. Clearly, this could also be a potentially critical single point of failure.

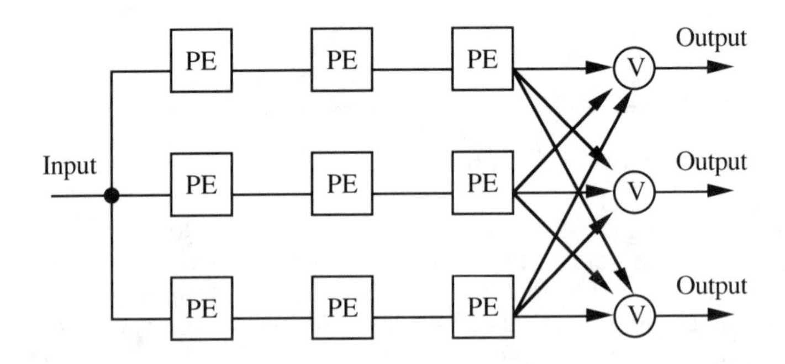

FIGURE 5.8 A TMR Linear Array with a Triplicated Output Voter

Considering the linear array multiprocessor approach described for this system, the location and frequency of the voting scheme may also be considered. While a single or triplicated voter located immediately prior to the computer system output can be effective in tolerating single permanent or transient hardware faults, voting could take place more frequently throughout the system. In this way, by dividing the array into k partitions, single permanent or transient hardware faults can be masked in each of these partitions (see Figure 5.9). This gives an overall improvement in fault coverage approaching k faults, assuming no more than one fault per partition. This technique can even be expanded to include voting on triplicated inputs, so that input errors are prevented from propagation through the system.

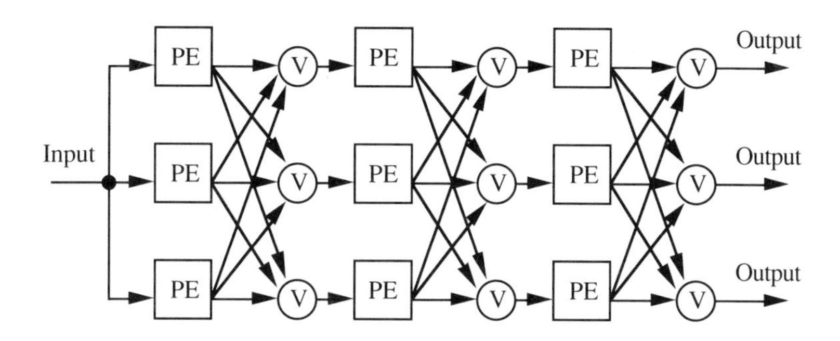

FIGURE 5.9 A TMR Linear Array with a Triplicated Voter at each Stage (k = 3)

However, these improvements come with an increase in hardware resources, in terms of both voters and connections. While hardware costs are not the primary concern in this TMR system, the attainment of the basic reliability requirements at the highest level of real-time performance is a singularly important goal. Therefore, since a single stage of voting at the system output is sufficient to meet the reliability requirements previously described, and since increased levels of voting would imply decreased real-time performance, use of multiple-stage voting will be reserved for future study.

5.5.1 Hardware-based vs. software-based voting

Now that the location and frequency of the voting process has been determined, the technique by which each individual voter is implemented must be considered. Voting can be employed using either hardware or software resources.

As implied in Figures 5.7-5.9, hardware-based voting consists of a group of parallel majority gates, implemented using a variety of combinational logic methods and synchronized using latches, which operate by sampling inputs in a bit-wise fashion and producing an output which represents a majority or quorum of the input values. Hardware voting is characterized by faster processing and less delay, but requires additional hardware resources. However, an underlying premise of TMR in general and this system in particular is that hardware costs are negligible in achieving overall system performance and reliability goals.

By contrast, software-based voting makes use of the processors in the system to perform the voting process, as shown in Figure 5.10. There are two problems which make software-based voting inappropriate for this system. First, the performance requirements for real-time DSP implementations dictate the need for faster processing and less delay. By having the voting process take place on the processors themselves, significant execution time is lost by both the voting process and the gathering of the data values themselves. Second, as discussed in section 5.8, hardware support is provided in each DSP96002 microprocessor for interprocessor communication with arbitration for multiple processors having access to a shared resource, such as another processor. This arbitration represents a single point of failure, since the failure of one processor (i.e. an insane node) during this arbitration process could potentially render all subsequent stages of all the linear arrays inoperable. Techniques can be used to offset this problem (e.g. dual-port memory between stages), but these approaches dictate hardware support. In addition, in order to provide a single output result, one of the processors must choose the output, thereby making that PE a single point of failure. This too can be avoided by the use of hardware support, but this even more tends to resemble a form of hardware-based voting.

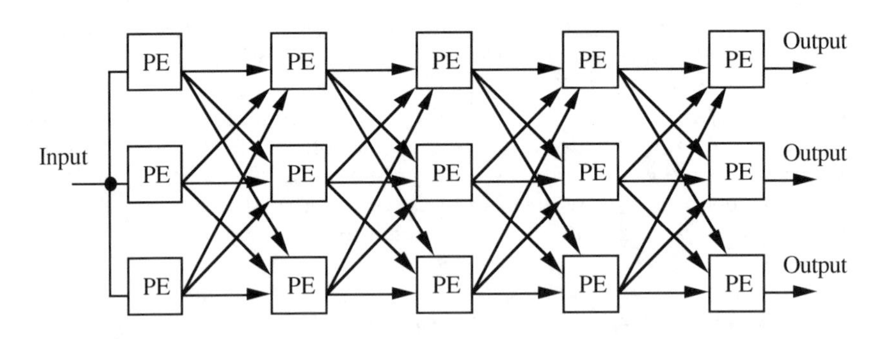

FIGURE 5.10 A TMR Linear Array with Software-based Voting

Johnson describes five decision rules for deciding between the use of hardware or software voting techniques [JOHN89a]. The choice depends on: the availability of one or more processors to perform the voting (if processors are available, software voting may be preferable); the speed required for voting (hardware voting provides more performance potential); the criticality of such issues as space, power, and weight limitations in the system implementation (software voting does not further burden these limitations); the number of voters required (more voters will imply more resources in terms of dedicated hardware or processor overhead); and the flexibility desired for voting to support future system upgrades and changes (software is typically more flexible than hardware). For example, if space, power, and weight limitations are paramount, software voting is clearly indicated.

Clearly, the critical tradeoff when choosing between hardware and software voting is one of execution speed versus hardware resources. Since real-time performance is of a primary concern in this system, the use of processor-based software voting would tend to degrade the capabilities of the system to perform DSP operations in a high-speed manner. In addition, the complexity of a single dedicated hardware voter stage is considered to be minimal with respect to the processors themselves, in terms of cost, space, power, and weight. Finally, the flexibility of the system with respect to future upgrades would not be compromised by hardware voting, since the voting hardware could be replicated at various stages throughout the architecture should the need arise.

Thus, when considering the issues presented in this subsection, hardware-based voting appears to be the appropriate choice for this system. A fault-tolerant voter will be used at the end of the TMR linear arrays to determine a single majority output for the system.

5.5.2 Single-output fault-tolerant voting

One final problem that is common to all fault-tolerant voting techniques is the method by which a single output is determined. In all the cases considered, the outputs themselves inevitably became triplicated, thereby requiring some form of voter or selector to determine a single output. This voter must incorporate a fault-tolerant design, since it clearly represents perhaps the most critical source of problems as a single point of failure.

Since the original TMR research implicit in [NEUM56], the majority voter for a TMR implementation has been an inherent single point of failure. Often this voter is assumed to be much simpler than the processors themselves, and thereby less prone to failure. Another approach is to design the voter using components and technology more reliable than that of the processors, thereby also providing a level of fault avoidance as opposed to fault tolerance. However, the most suitable solution to this problem is to design the voter so that its design incorporates fault-tolerant operation.

The technique that will be used in this system is adapted from a design described in [WENS85]. The circuit described by Wensley for fault-tolerant majority voting on single-bit inputs producing a single-bit output is shown in Figure 5.11. Given correct inputs, the reader can verify by inspection that each individual active component in this circuit can experience a fault, modeled as either a stuck-at-0 or stuck-at-1 fault, without corrupting the output. Passive elements, such as the pull-down resistor and the wires, are not considered to pose a significant risk of failure.

Alternately, assuming correct operation of the active components in the circuit, the output will represent the majority value of the inputs despite an error on any single input. Since the data inputs to be received by the single-output fault-tolerant voter for this system are 32-bits wide, this circuit can be replicated in order to accommodate 32-bit data input voting and produce a 32-bit data output.

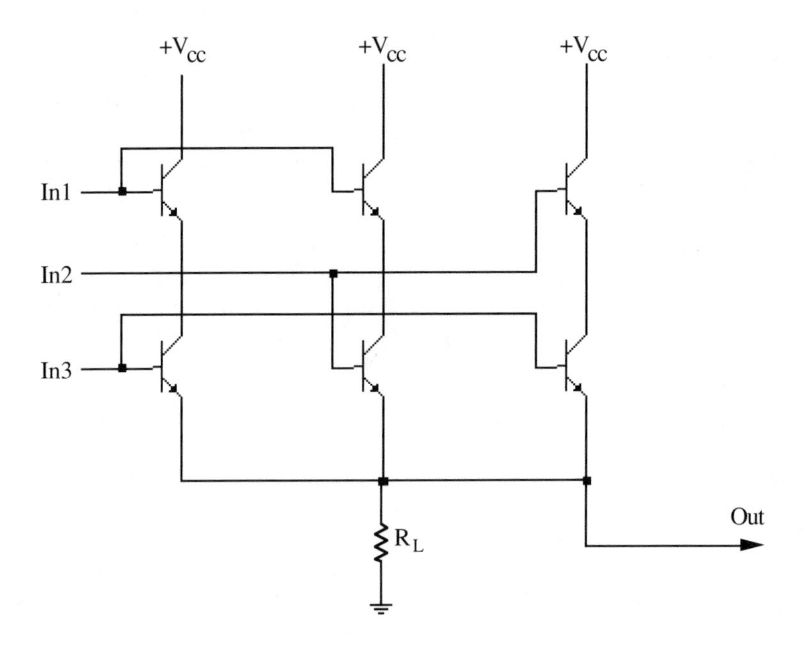

FIGURE 5.11 Single-bit Fault-Tolerant Majority Voter

5.5.3 Input selection

Another fundamental problem of voting is determining whether two inputs represent the same value. Since replicated real-time DSP system inputs derived from devices such as A/D converters often slightly differ in one or more least significant bits, it becomes crucial for the voter to be able to recognize their relationship.

One approach that has been used successfully is to consider the three TMR inputs in numerical order and select the middle value. When at most one input is in error, this results in one of the two correct inputs always being selected. Another approach consists of ignoring some predetermined number of least significant bits, so that the more important higher bits are used for the selection. Of course, it can be a potential difficulty to determine just how many bits to ignore. Too many ignored can lead to an invalid conclusion of equality, and too few can lead to a mistaken conclusion of differing values.

Since one of the advantages of the IEEE 754-1985 binary floating-point arithmetic standard used on the DSP96002 microprocessors is that the results are consistent across multiple executions, the only location where this value selection problem could potentially occur is at the inputs themselves. The application itself will determine whether or not input voting is necessary, depending upon whether triplicated input peripherals are used and if so, whether they have the potential to deliver differing values (e.g. a set of triplicated A/D converters could pose a problem).

The design of this system will support both singular and triplicated input configurations, although the former will often be used for examples since it is a mere combination of the latter. The triplicated input form enables the system to directly connect to input devices which produce triplicated outputs.

For example, in a data acquisition system, three A/D converters would be used. Each of these converters is fed by a common analog input signal and drives one of the inputs to the system. In order to overcome the input selection problem, some number of least-significant bits on these inputs would be tied to a constant zero. The exact number of least-significant bits to be truncated will of course depend on the specifications of the A/D converters themselves. In any case, the value selection problem is not relevant anywhere other than at the inputs to this system, due to the IEEE 754-1985 compliance of the DSP96002 devices.

5.6 Fault-Tolerant Clock Synchronization

To this point, the design of the system has included a separate and independent clock circuit for each PE. This provides a solution to the problem of fault-tolerant global clock synchronization in a multiprocessor system. However, it is possible to adapt this approach so as to improve system simplicity, predictability, and performance without sacrificing single-fault reliability.

Consider a TMR linear array system with three arrays of R stages each. Should any of the R independent clocks fail in a linear array, that entire array would potentially be rendered inoperable. Thus, even though the clocks in each array have been replicated, this increase in complexity provides no additional reliability.

However, by modifying the system design to include separate and independent clocks for each linear array and *not* each individual PE, a number of advantages can be attained. The system becomes simplified, both from a hardware and a software standpoint, since the processors in each array can operate in a synchronized manner like that of a systolic linear array. This in turn tends to improve the predictability and performance of real-time execution, since synchronization primitives between processors in each linear array become much simpler. And, since the failure of a single array clock will still result in a potential loss of only that array, the level of reliability of the system is not compromised.

One critical concern in any fault-tolerant multiprocessor system is the manner with which processors are synchronized. The need for synchronization in TMR-based systems is well understood, since the voting process itself is based on the premise that all input data arrives simultaneously [MEYE71]. However, at the same time, this synchronization can potentially form a major single point of failure. An overview description of fault-tolerant clock synchronization issues can be found in [RAMA90].

Synchronization techniques for redundant systems such as the one under consideration can be divided into three categories: independent accurate time bases; common external reference; and mutual feedback [DAVI78]. With independent accurate time bases, the processors are completely decoupled but can maintain synchronization for a short period of time. With a common external reference, the processors rely on a common external time base for synchronization. Finally, with mutual feedback, no common time reference is used, relying instead on mutual monitoring among processors for timing and drift corrections.

As noted by Davies and Wakerly, independent accurate time bases are impractical for most applications, since the periods with which they can maintain synchronization are brief [DAVI78]. Mutual feedback has not been used extensively to date, and is impractical for this system, since it implies a relatively high level of processor overhead. The appropriate choice for this fault-tolerant real-time DSP system is hardware-based implementation of the common external reference approach, since it eliminates processor overhead and is also the most widely studied, tested, and used.

The types of faults that must be considered in the design of a fault-tolerant common reference clock include both the classical stuck-at-1 and stuck-at-0 faults, as well as *malicious* failures [VASA88]. Malicious failures are those clock failures that are perceived differently by different non-faulty clocks. In general, if the failure of f clock modules is to be tolerated, then the clock system must incorporate at least $(3f + 1)$ clock modules [DALY73, KRIS85]. Therefore, based on the reliability requirements previously enumerated, a minimum of four clock modules will be needed in the design of the fault-tolerant clock portion of this system.

Hardware implementations of the fault-tolerant common external reference clock use the principle of phase-locked loops in order to achieve tight synchronization between clocks which have been replicated to provide a level of fault tolerance [SHIN88]. This synchronization is provided with no appreciable time overhead, assuming that transmission delays in the wires themselves are negligible. This assumption will be made here, since the physical distance between clocks is not expected to be substantial.

The design of the fault-tolerant global clock for this system will be based on these phase-locked loop techniques as outlined in studies such as [VASA88, KRIS85, SHIN87, SHIN88, SMIT81]. A block diagram description of the clock design is shown in Figure 5.12.

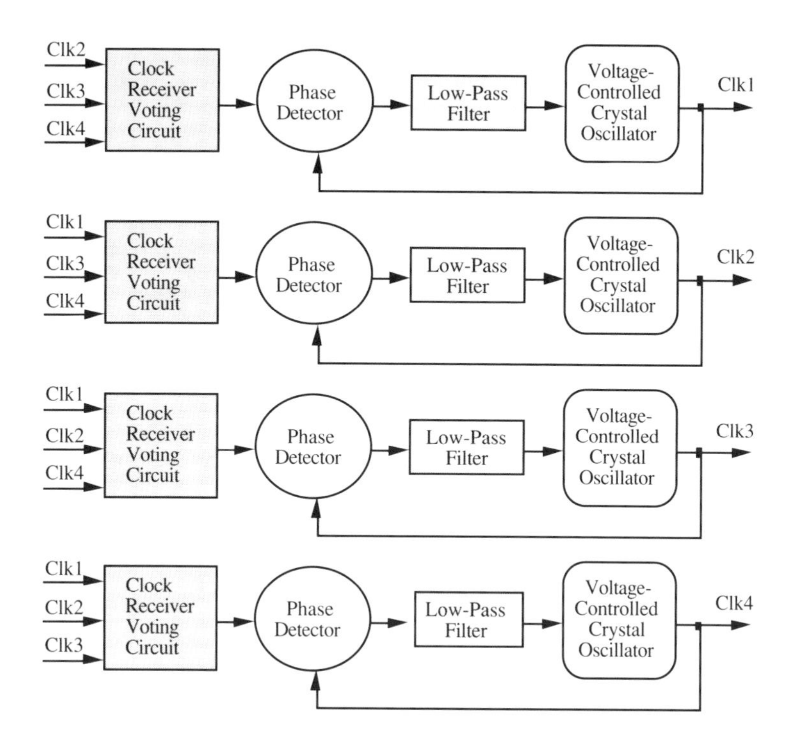

FIGURE 5.12 Phase-Locked Clock Design

The basic operating procedure carried out by the phase-locked loop clock modules is relatively straightforward. Each of the four clocks is in itself the output from a voltage-controlled crystal oscillator. The voltage applied to this oscillator is determined by the filtered output of a comparator or phase detector, which is proportional to the difference between the output of the clock it drives and a reference clock determined from the median of the other three clocks.

Due to the TMR orientation of the fault-tolerant real-time DSP system being developed, the processors themselves can naturally be divided into three groups based on which linear array the processor appears in. Of the four clock outputs emerging from this fault-tolerant clock synchronization module, one will drive the processors in the first array, another the processors in the second array, and another those in the third and final array. The fourth clock output will be held in reserve for future use. A 3-stage TMR linear array system using these clocks is shown in Figure 5.13.

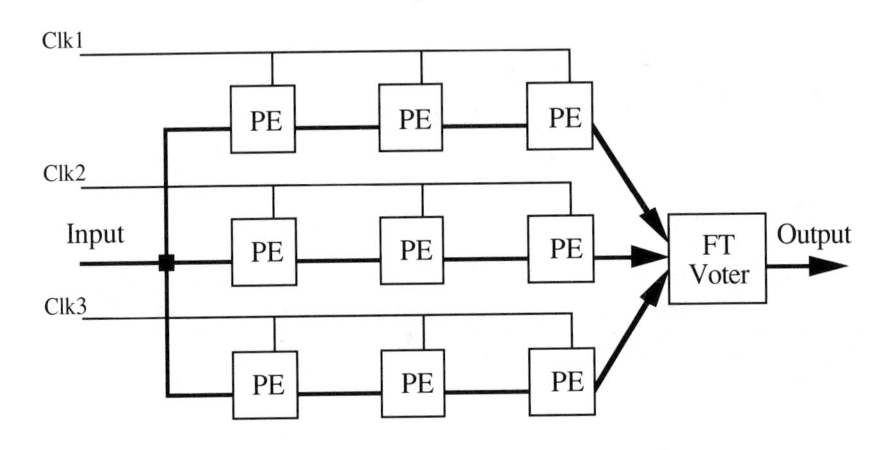

FIGURE 5.13 A 3-stage TMR Linear Array System

Although all processors operate using a globally-synchronized clock, communication between processors in each array will be asynchronous. In this way, while the three processors in each stage operate in a lock-step fashion, those processors in different stages will not. This will provide the system with an added level of flexibility, since the application software to execute on each stage of processors can be developed without the problem of complex lock-step timing between stages. Thus, the task performed by a given stage can be altered without major modification to the tasks on other stages.

5.7 Communication Primitives

The system design consists of a TMR linear array of PEs where each PE consists of a DSP96002 microprocessor and local memory. An important consideration in the design of this system is the operating system. Since this system is a loosely-coupled multiprocessor or distributed processing system, a distributed operating system could be considered. However, since this system is more specialized than the general-purpose multi-user systems normally associated with distributed operating systems, only some of the same issues can be considered and applied.

As discussed in section 3.3, the issues associated with the implementation of distributed operating systems can be divided into five categories. Of these, only the communication primitives are relevant in a special-purpose system such as this one. The important considerations for these message passing primitives are reliable transmission, blocking, and buffering operations.

The operating system for this fault-tolerant real-time DSP system will use unreliable communication primitives to minimize delays, since the communication lines between processors are minimal and unlikely to fail. Even if they do fail, the fault tolerance provided by the static redundancy scheme would still prevail. These primitives will be blocking primitives, since this makes the communications simpler and much more straightforward with respect to time-critical operations such as those in a real-time DSP system. Finally, the primitives will be single-element buffered, using hardware features provided by the DSP96002 microprocessor, so that delays can be reduced.

One key consideration in the design of the operating system and its communication primitives for this system will be the hardware support provided by the DSP96002 microprocessor. This support will make it possible for the primitives to operate with maximum efficiency, thereby increasing the real-time DSP performance of the system. These and other issues will be addressed in the next two sections.

5.8 System Interface and Communication Design

Although the basic system design has been developed, many details remain to be considered. In particular, a method for communication between microprocessors and other components must be developed.

5.8.1 Memory interface model

Since both the DSP96002 and its HI are based on memory-mapped I/O communication, a general memory interface will first be designed. This will then serve as a model for subsequent PE-to-PE, PE-to-output, and input-to-PE hardware and software interface designs. The basic design of this interface model is shown in Figure 5.14.

Pins S1 and S0 determine which address space (X, Y, or Program) is requested, so that S1, S0, and the address bus (A0-A31) together determine the complete address and are stable when TS* is asserted. The address decoders use these pins as inputs to formulate whether or not the memory device or devices being driven are appropriate for the address access requested. The type of access (read or write) is of course determined by each R/W* output pin. The TA* (Transfer Acknowledge) pins are used as feedback lines for adding wait states, and have been tied to logic-0 to indicate that the SRAMs (static random-access memories) are fast enough to operate without wait states.

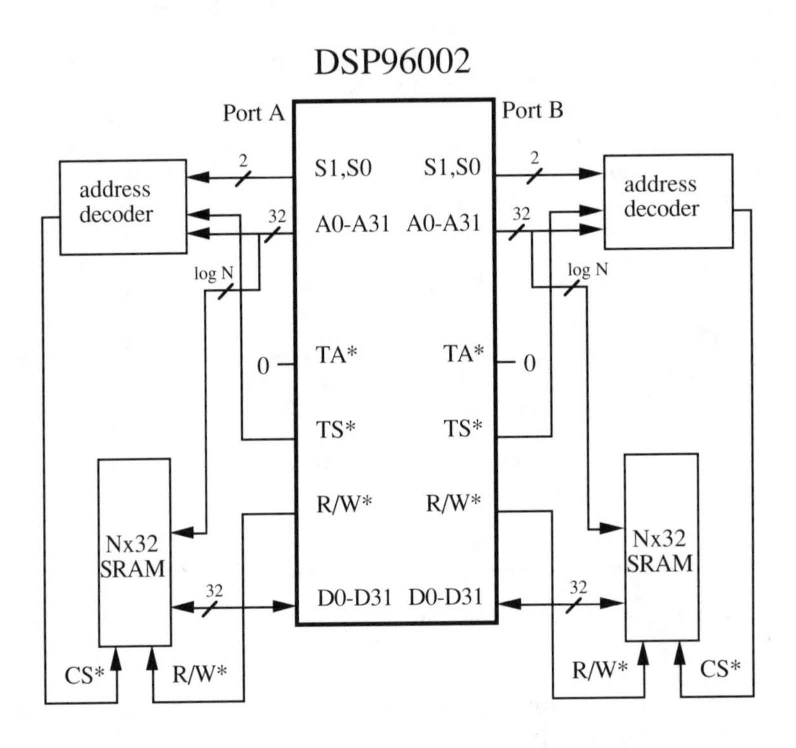

FIGURE 5.14 DSP96002 Memory Interface Design

In order to determine where each external memory address should be mapped by the microprocessor (i.e. whether to port A or B), the PSR (Port Select Register) is used. Each of the address spaces (X, Y, and Program) are divided into eight sections, and a bit is provided in the PSR for each of these twenty-four total sections, where a zero indicates a mapping to port A and a one to port B.

By using the general external memory interface designed for the microprocessors in this system as a model, the designs for PE-to-PE, PE-to-output, and input-to-PE interfaces can be devised. These will be considered individually.

5.8.2 PE-to-PE interfacing

Each source or master PE in the TMR linear array system will communicate with its destination or slave in the array via memory-mapped access on the master PE and HI access on the slave. The master DSP96002 will use its port B for this communication, while the slave will use its port A.

As a general rule, the external X address space of every DSP96002 in this system will be mapped to port A and the external Y address space will be mapped to port B. In this way, the master processor will write information to the slave using a predefined Y memory address which represents (via its A2-A5 values) a *TX register write* HI function call. Similarly, the master processor will read feedback status information from the slave (e.g. to determine if the TX register on the slave's HI is empty) using another predefined Y memory address which represents an *ICS register read* function call.

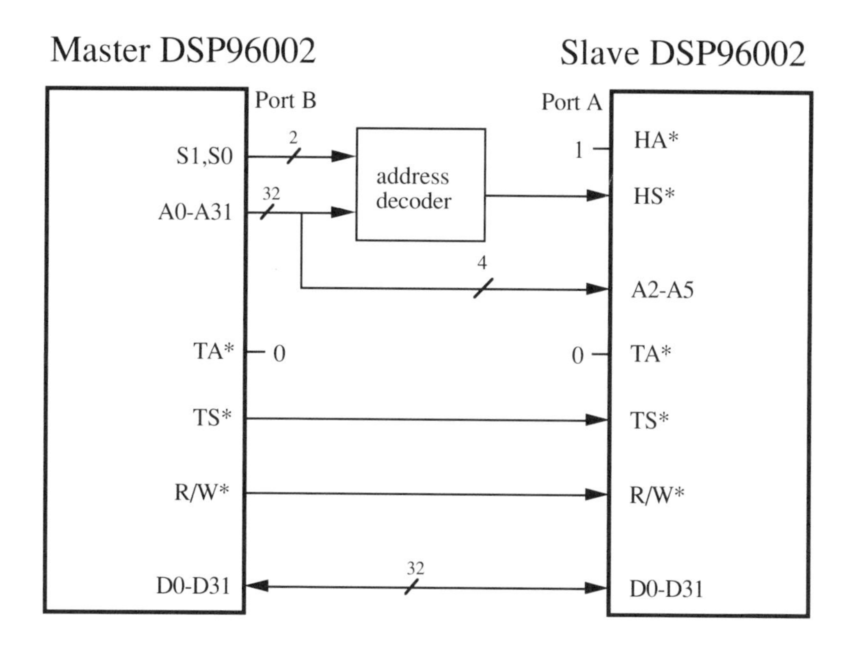

FIGURE 5.15 PE-to-PE Interface Design

These memory-mapped addresses will, along with S1 and S0, emerge from the master processor and serve as inputs to an address decoder. This decoder will check the complete address requested (including S1 and S0) and assert HS* on the slave processor if and only the address is one of the two predefined Y addresses described above. This in turn will enable the HI of port A on the slave processor and thereby instigate the interprocessor communication. A block diagram description of this PE-to-PE interface design is shown in Figure 5.15.

The PE-to-PE interface software will have the following basic structure. When the master has data to be sent to the slave, it must first wait until the TX register on the slave is free. This can be done by polling the appropriate bit in the slave's ICS register using the *ICS register read* HI function, or via interrupts with or without DMA. When TX becomes available, the master sends the data to the slave using the *TX register write* HI function.

Similarly on the slave, the HI interface hardware manages the double-buffer system. That is, when the TX register receives a value, the HI hardware moves this value to the HRX register when the latter becomes free. From a software standpoint, when the slave desires input data, it must first wait until the HRX register is full. This can be done by polling the appropriate bit in the HSR register, or via interrupts with or without DMA. When HRX is filled, the slave reads the data and can then begin to process the information.

5.8.3 PE-to-output interfacing

The interface between a PE and an output (voter) is very similar to the interface of a single-element memory device to a processor. A 32-bit latch is used to provide stable data for the fault-tolerant voter. The DSP96002 is the master, and communicates with the latch over its port B using memory-mapped access (not the HI) and a predefined address at the high end of the Y address space. As with the PE-to-PE interface design, an address decoder enables access to the latch if and only if the predefined address is requested. This address (to be denoted SYS_OUT) is write-only by the DSP96002 and read-only by the voter. A block diagram description of the PE-to-output interface design is shown in Figure 5.16.

Synchronization of output data is provided by an output ready strobe on each final-stage processor. All three together are used to drive a fault-tolerant majority gate whose output indicates when the voted output data is available for access. Transfers from the processor to the latch do not need to be synchronized, as any output devices are assumed to be capable of reading the present voted output data before the linear processor array and fault-tolerant voter can produce a new output.

The PE-to-output interface software has the following basic structure. When the processor is ready to output data for voting, it writes the data to memory location Y:SYS_OUT. After writing to this location, the processor is free to continue the processing of another input value to produce a new output result.

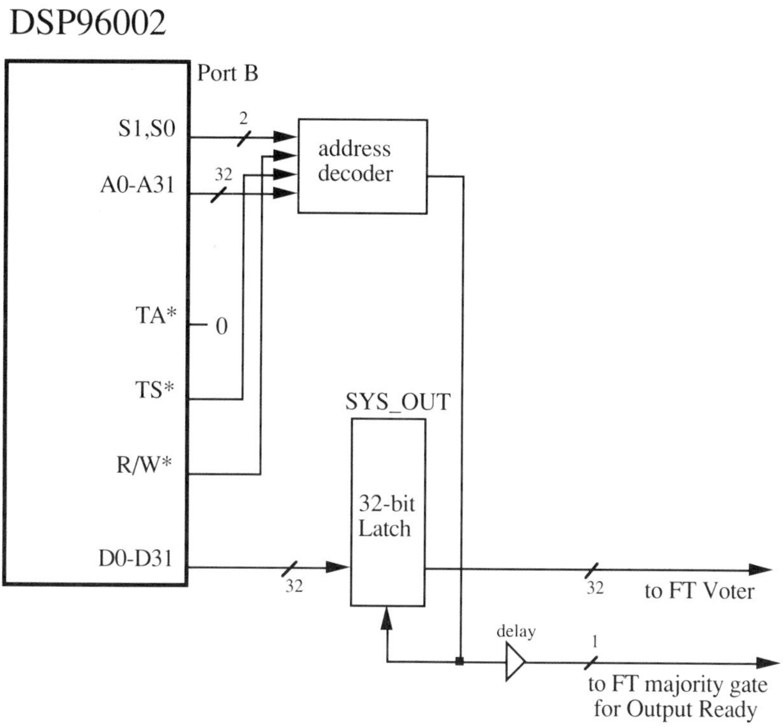

DSP96002

Port B

S1,S0

A0-A31

TA*

TS*

R/W*

D0-D31

2

32

0

address decoder

SYS_OUT

32-bit Latch

32

32

to FT Voter

delay

1

to FT majority gate for Output Ready

FIGURE 5.16 PE-to-Output Interface Design with Voter

5.8.4 Input-to-PE interfacing

The interface between an input (either the system input or a previous voter) and a PE is very similar to the interface of a single-element memory device to a processor. One 32-bit latch is used to hold stable input data. The DSP96002 is the master, and communicates with the latch over its port A using memory-mapped access (not the HI) of a predefined address at the high end of the X address space (to be denoted SYS_IN). As with the PE-to-PE interface design, an address decoder enables access to the latch if and only if the predefined address is requested. The latch is write-only by the input or voter and read-only by the processor. A block diagram description of the input-to-PE interface design is shown in Figure 5.17.

Synchronization of input data with the latch is provided by the input ready strobe, which indicates when the latch has been read by the processor and is available to receive new data. Transfers from the latch to the processor do not need to be synchronized, as any input drivers are assumed to be capable of writing new input data to the latch before the linear processor array can request a new input.

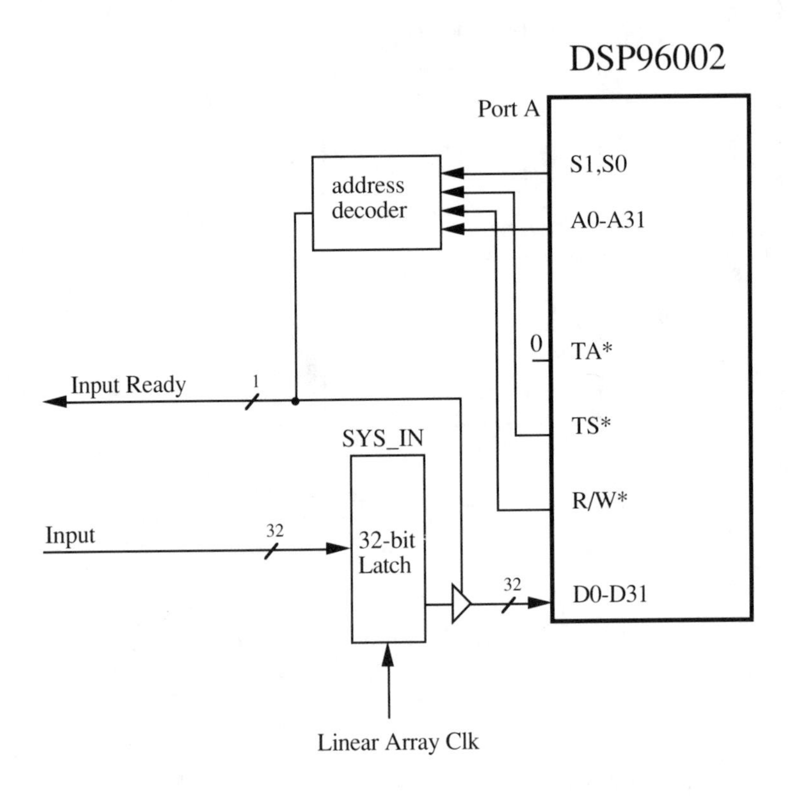

FIGURE 5.17 Input-to-PE Interface Design

The input-to-PE interface software has the following basic structure. When the processor is ready to input data for processing, it reads the data from memory location X:SYS_IN. The latch is updated using the same clock as the processor, so that the processor can access the most recent input data in the latch without any need for arbitration. A block diagram description of the communication flow in a 3-stage TMR linear array system is shown in Figure 5.18 (where L = latch).

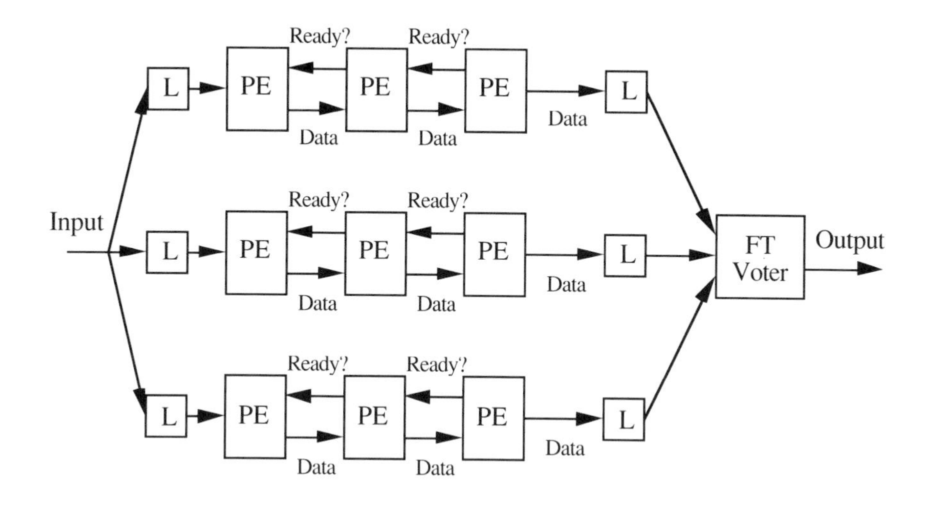

FIGURE 5.18 Communication Flow in a 3-stage TMR Linear Array System

5.9 Design Enhancements for a Dual-Mode Architecture

Since some applications require higher levels of reliability than others, and in fact applications can require higher levels for only a critical section of execution, the system design can be enhanced to provide a dual-mode architecture. The two modes of this architecture are FT (fault-tolerant) and non-FT. The FT mode consists of the TMR linear array system designed to this point, while the non-FT mode uses all processors to form a single *superarray* three times the length of the FT mode configuration, as shown in Figure 5.19. This provides a potential performance improvement approaching threefold over that of the FT mode configuration, with of course a substantial sacrifice in reliability.

The choice of either FT or non-FT operational mode will be one determined by the software executing on the individual processors. Hardware support is included for both modes, so that a particular application can make a transition from one mode to the other as necessary, with little or no overhead incurred.

As Figure 5.19 illustrates, the changes that are required to provide support for the non-FT mode simply consist of using combinations of two of the interfacing techniques previously developed. The changes manifest themselves only in the inputs to the initial stage of processors and the outputs from the final stage of processors. The final-stage processors combine both the PE-to-output and PE-to-PE interface designs, while the initial-stage processors combine both the input-to-PE and PE-to-PE interface designs.

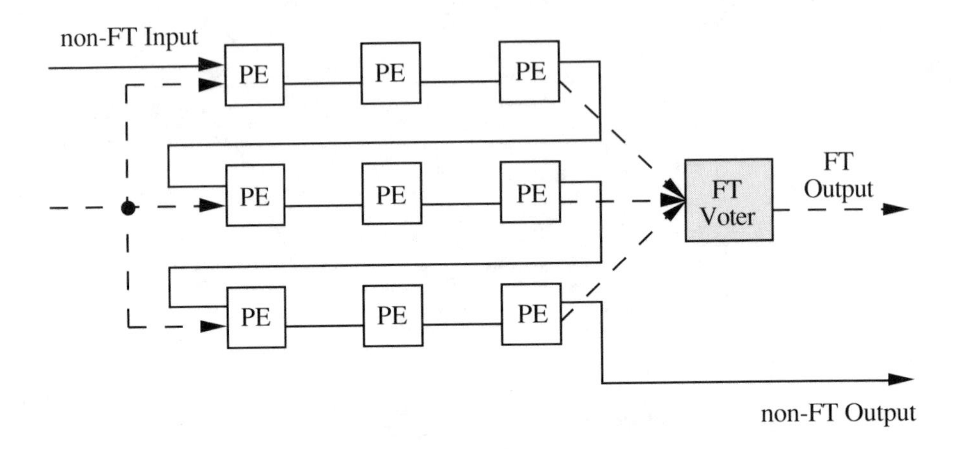

FIGURE 5.19 A 9-PE Dual-Mode System in non-FT Mode

5.9.1 Dual-mode final-stage output interface design

On the three final-stage processors in the dual-mode system design, there are two different interface designs used. One design is used with the final stage processors in the first and second linear arrays, while another is used with the final-stage processor in the third and final array.

The output interfacing design for the final-stage processors in the first and second arrays consists of a combination of both the PE-to-output and PE-to-PE interface designs, as shown in Figure 5.20. As previously developed, the PE-to-output connections consist of driving a latch which in turn drives the fault-tolerant voter. As the voter receives its inputs, it outputs the voted result, and a fault-tolerant majority gate is used to determine the FT mode output ready strobe. The PE-to-PE connections consist of selecting HS* on the slave destination processor when either an *ICS register read* or *TX register write* HI function is requested, along with direct connections of the D0-D31, A2-A5, R/W*, and TS* pins.

The design for the final-stage processor in the third and final array consists of just a PE-to-output interface along with support for the superarray non-FT output, as shown in Figure 5.21. The only additions to the FT mode output design are the non-FT output (without voting) along with an output ready strobe to indicate that the superarray output data is available.

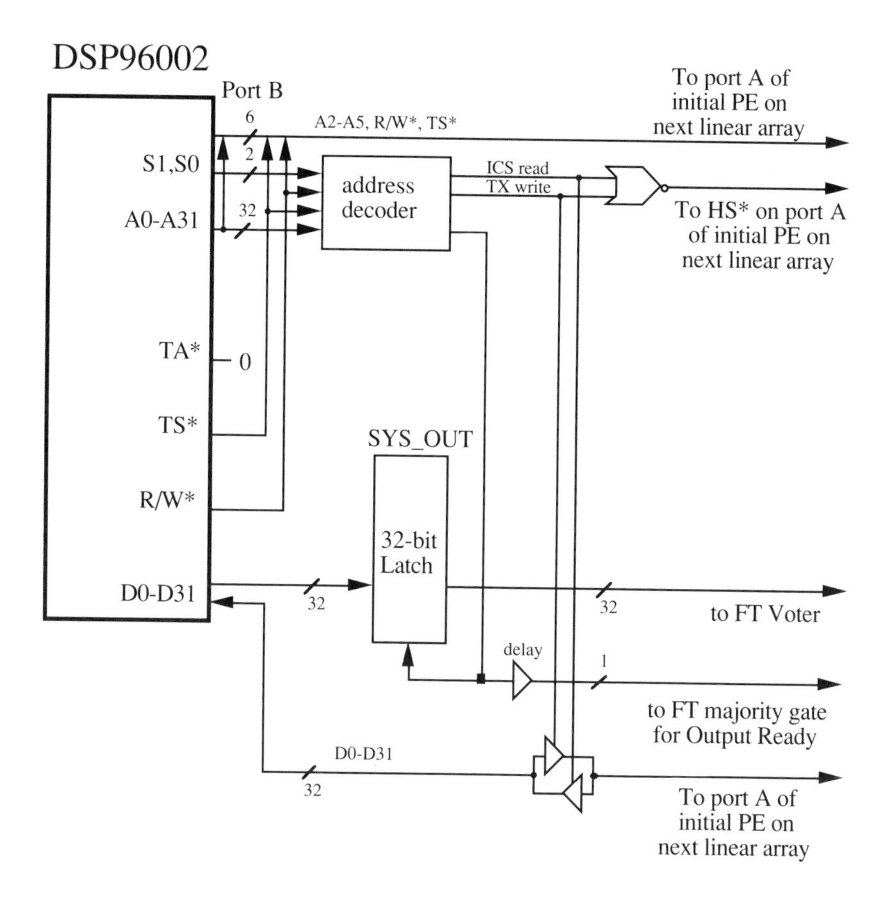

FIGURE 5.20 Dual-Mode Final Stage Output Interface Design (first and second linear arrays)

The interface software which will run on all three final-stage processors will perform different functions depending on the desired mode of operation. In FT mode, as before, all three processors write their output results to Y:SYS_OUT in their own address space. The non-FT output and strobe lines remain unused.

In non-FT mode, the final-stage processors in the first and second linear arrays each communicate with their respective destination processor in the initial stage of the next array to form the superarray, using *ICS register read* and *TX register write* HI function calls as previously described with PE-to-PE communication. The final-stage processor in the third array, however, forms the superarray system output by writing to Y:SYS_OUT. The fault-tolerant voter, the fault-tolerant output strobe majority gate, and their input and output lines all remain unused.

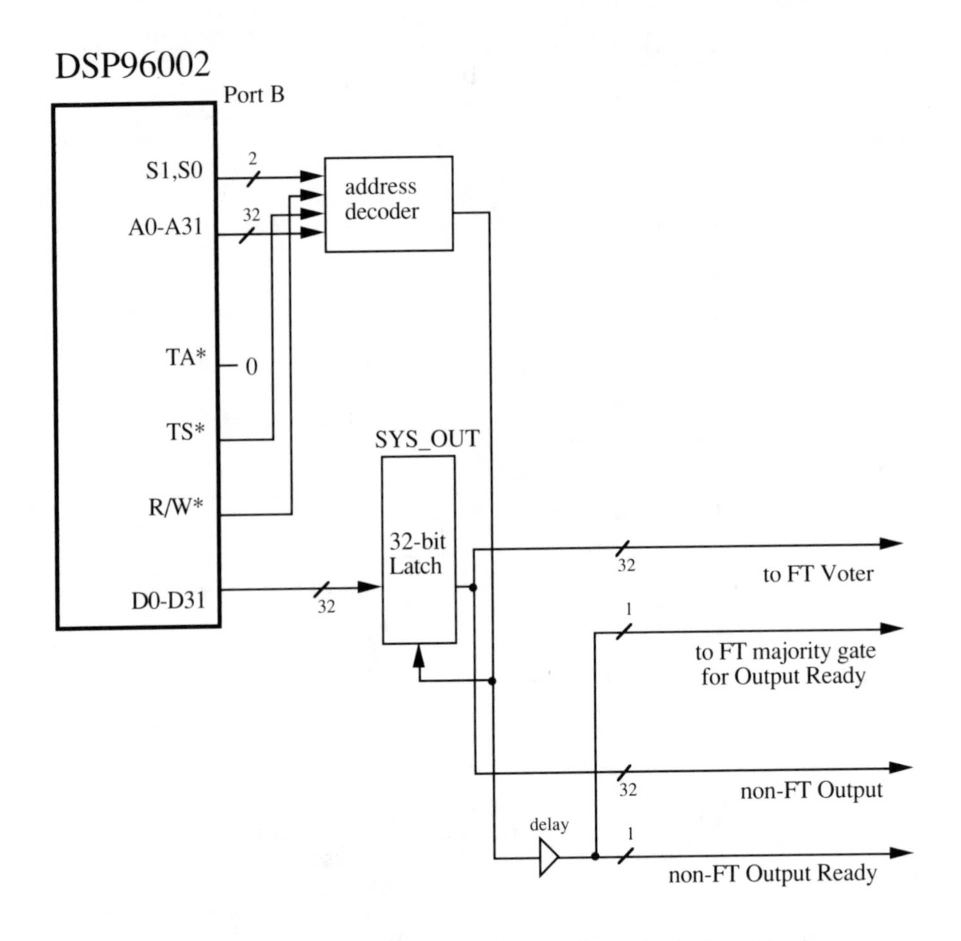

FIGURE 5.21 Dual-Mode Final Stage Output Interface Design (third linear array)

5.9.2 Dual-mode initial-stage input interface design

In order to support dual-mode operation, the interface design of the initial-stage processors must be modified. The design for the initial-stage processor in the first linear array is the same as before, except that its input serves as both *a* FT mode system input and *the* non-FT mode superarray system input. The processors in the second and third arrays, however, must combine both input-to-PE and PE-to-PE communication. The new connections necessary to support the additional role of PE-to-PE communication were described in the dual-mode final-stage output design discussion earlier in this section.

The interface software which will run on all three initial-stage processors also will perform different functions depending on the desired mode of operation. In FT mode, as before, all three processors read from X:SYS_IN in their own address space. In non-FT mode, the initial-stage processor in the first linear array reads from X:SYS_IN to retrieve each non-FT input value. The initial-stage processors in the second and third arrays, however, each read from their predecessor in the single non-FT superarray. The method with which they read these PE-to-PE values is identical to that described for the slave processors in the PE-to-PE interface discussion.

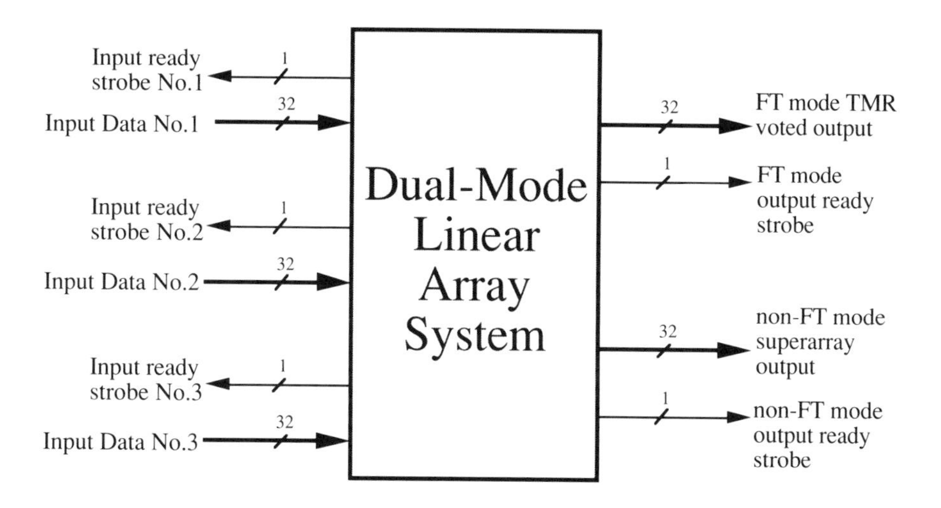

FIGURE 5.22 External Input and Output Connections

With these designs completed, the system is ready to be simulated, with external system input and output connections as shown in Figure 5.22. The system design consists of a linear array multiprocessor system with two modes of operation. In FT mode, the system behaves in a TMR fashion with fault-tolerant array clock synchronization and voting. This mode was chosen to provide single-fault tolerance of virtually all permanent and transient hardware failures with no degradation in real-time DSP performance. In non-FT mode, the system behaves in a *superarray* fashion, thereby increasing potential levels of performance at the expense of fault tolerance. The selection of the mode itself is determined by the communication interface instructions used on each PE, thereby enabling the system to make a transition from one mode to the other as run-time application requirements dictate. These instructions will be based on simple operating system communication primitives to provide a level of software fault avoidance.

5.10 Summary

This chapter has presented a careful analysis of the goals and premises for this system and then, based on these goals and premises, has outlined the design process for the entire system.

The first design issue was that of real-time vs non-real-time support. The decision was made in favor of real-time since it is simpler to modify a real-time system to non-real-time applications than vice versa.

The input and output for the system reflected a simple data acquisition model. With this model, input data is sampled at a rate greater than or equal to the capabilities of the system, the data is processed by the system, and the results are produced as output.

In order to best determine an overall architecture for the system, potential parallel architectures for DSP applications were then considered. After a careful analysis of DSP requirements in terms of the representative operations chosen (i.e. digital filtering and fast Fourier transforms), it became clear that an MIMD linear array of microprocessors was the most desirable parallel system.

The parallel architecture model used by this system is basically an MIMD processor array or grid architecture using a linear interconnection pattern between neighboring nodes. This system has the ability to support multiple instruction streams and multiple data streams flowing in either a unidirectional or bidirectional fashion. As is often the case with processor array architectures, a particular interconnection pattern was chosen for this system to exploit the concurrency of the algorithms to be used.

Of the two basic approaches to processor arrays (i.e. systolic arrays and wavefront arrays), the wavefront approach is perhaps the more suitable model since, like wavefront arrays, this system uses handshaking protocols to provide asynchronous interprocessor communication and operations. However, one important distinction between this system and the common methods of both systolic and wavefront arrays is the size and power of the individual nodes and the granularity of parallelism that is employed. While systolic and wavefront arrays are associated with fine-grained machines consisting of extremely small and simple PEs, this system instead uses a more medium-grained level of concurrency with advanced and extremely complex microprocessors serving as nodes.

This method could be described as starting with the best in DSP uniprocessors as the basis for designing a DSP multiprocessor. The use of other such advanced microprocessors has also been the basis of multiprocessor architectures like the Intel iPSC/1, iPSC/2, and iPSC/860, the Cal Tech Cosmic Cube systems, and the Thinking Machines CM-5. However, in contrast, all of these systems incorporate a hypercube interconnection structure.

FT goals of DSP were considered next, with emphasis on high-availability, long-life and critical-computation goals. The last of these three goals was found to be the most important for real-time DSP applications, and this resulted in most of the FT emphasis being centered on a high level of reliability. As before, the particular representative DSP operations used were digital filtering and fast Fourier transformations, where reliability and performance needed to be dynamically balanced along with cost considerations.

Simplicity of design and the use of reliable components argued for the use of off-the-shelf hardware devices. This also implied that sharing of elements should be minimized.

Based on real-time constraints, the system was required to tolerate any single permanent or transient hardware fault. The sources of failures considered were: processor and memory failure; communication network failure; peripheral device failure; environmental and power failure; and human error. The first two of these categories were given the most emphasis since solutions for the remainder had been addressed in the literature.

Static hardware redundancy via TMR was selected as the most appropriate redundancy approach for this system, as was a linear array configuration for performance reasons. The seven basic categories of fault-tolerant multiprocessor communication architectures were considered, but as they all were either designed for dynamic hardware redundancy or had other incompatibilities, they were rejected for the system. However, the triple modular redundancy architecture, being a static hardware redundancy technique which balanced reliability and performance with the low cost of off-the-shelf components, satisfied the stated requirements.

In evaluating this design, a major consideration was the issue of the voter, since if it failed, the system failed. Triplicating the voter addressed this problem, but as eventually a single output was needed, an inherently FT single-output voting approach was needed. Both hardware and software voters were considered. The former provided faster processing and less delay, but demanded extra hardware. On the other hand, the software approach used system processors, but since DSP applications required fast processing and communication arbitration could cause failures, this approach proved unacceptable. Moreover, as hardware costs were considered negligible, the hardware voter was selected.

The actual implementation involved FT majority voting on single-bit inputs, producing a single-bit output. Thus each component could have a stuck-at-0 or stuck-at-1 fault without corrupting the output. This was replicated for the 32-bit wide data. Techniques for input voting were discussed, since the input itself could also require voting, however this was considered application dependent.

Rather than a single clock per PE, a clock per linear array was proposed, since if a clock on one PE failed then so did the whole array. A clock per array simplified the system by simplifying synchronization primitives in the array without any loss of reliability. These clocks operated using a phase-locked technique providing a globally-synchronized clock, allowing processors in each stage to operate in a lock-step fashion without restricting processors in different stages.

The operating system design used unreliable communication primitives to minimize delays since communication lines between lines were minimal and hence unlikely to fail. However, if they were to fail, the static redundancy scheme would prevail. Communication was made simpler by making the primitives blocking and single-element buffered using hardware features available on the DSP96002. These features allowed for the increased performance needed in real-time applications.

This completed the basic system design. The remainder of the chapter dealt with more detailed designs including communication interfaces between processors and other components. In particular, the host interfaces, memory interface, PE-to-PE interfacing, PE-to-output interfacing, and input-to-PE interfacing were presented in much detail.

The final interface considered allows for a unique feature of the design. As some applications require higher levels of reliability than others, or indeed critical sections require higher reliability, the system was enhanced to provide a dual-mode architecture (i.e. FT and non-FT). The former was the TMR linear array as described, whereas the latter used all the processors in a single superarray three times the length of the FT mode architecture. This gave a potential three-fold performance improvement over the FT mode, but with a significant decrease in reliability. The choice between modes was to be a software decision; hardware for both modes was included and the transition between modes was to be made with little or no overhead. In particular, the changes that would occur would be in the input to the initial stage of processors and in the output from the final stage of processors. Final-stage processors combined PE-to-output and PE-to-PE interface designs, while the initial-stage processors combined both input-to-PE and PE-to-PE interface designs.

Now that the design has been completed, the next step is the development of the simulation software for this system. Many issues will be considered in order to simulate this system, such as simulation library support, floating-point conversion, voting, PE-to-PE communication, PE-to-output communication, and input-to-PE communication.

Chapter 6
SYSTEM SIMULATION

This chapter describes the development of the simulation software for the system. Many issues are considered in order to simulate this system, including simulation library support, floating-point conversion, voting, PE-to-PE communication, PE-to-output communication, and input-to-PE communication.

6.1 Simulation Library Support

The DSP96002 simulation software consists of a series of object libraries which support one of two possible modes of execution. A display version of the simulator includes a wide variety of functions for screen management and interactive debugging aids, while a non-display version is more streamlined and takes approximately half as much memory.

The display version is more suitable for single processor simulation, while the non-display version provides the designer with more flexibility in the simulation of multiprocessor systems. Therefore, in order to support larger system simulations with greater flexibility, the non-display mode of operation will be the primary emphasis in the simulation of this system. However certain issues, such as the initial testing of application and operating system software on a single processor, will tend to demand the use of the display version on occasion.

The non-display version of the DSP96002 microprocessor simulation object libraries consists of twenty-one C functions which are callable from user programs [MOTO89]. These functions are described in Appendix A.

In order to simulate a working multiprocessor, each processor is first created using the *fsp_new* function. This function allocates and initializes the device state data structure that will completely describe the particular DSP96002 device throughout its execution. By using a device index, the *fsp_exec* function enables an individual processor to execute a single clock cycle, so that simulation of parallel execution can be achieved by iteratively calling *fsp_exec* for each processor in turn. Other functions provide low-level support for accessing individual DSP96002 pins, ports, registers, and memory locations. The remaining functions are provided for simulation support, such as the ability to save the state of the processors to a set of files for later recall.

6.2 IEEE Single-Precision Conversion

One preliminary consideration that must be addressed before system simulation can be developed is the need for data display and evaluation. While floating-point values are carried internally throughout the system, and are manipulated in the microprocessors, these values are passed between components in IEEE 754-1985 single-precision (SP) floating-point format [IEEE85]. This format is not the same as that used by either *float* or *double* floating-point variables in the C compiler used to simulate this system. In fact, these IEEE SP values are stored as 32-bit *unsigned long integers* in the simulation software. Therefore, the need exists for functions in C to convert between these *unsigned long integers* representing IEEE SP and the *double* format of the compiler so that the flow of information in the system can be better understood.

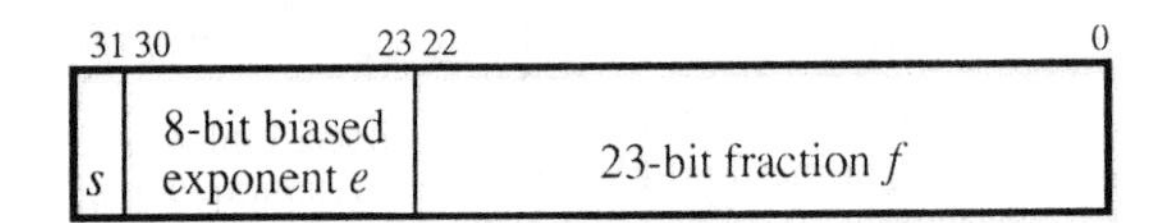

FIGURE 6.1 IEEE 754-1985 Single-Precision Format

The basic format of IEEE SP contains three fields, as shown in Figure 6.1. Twenty-three bits of binary fractional data and eight bits of biased exponent (as a power of two) are provided, along with a sign bit. Normalized nonzero numbers x are equal to the following, where s represents the sign bit, e the exponent, and f the binary fixed-point fraction $0.b_1b_2...b_{23}$ [DSP89]:

$$x = (-1)^s \times 2^{e-127} \times 1.f \tag{6.1}$$

$$1.f = 1 + (0.5) \times b_1 + (0.25)b_2 + \ldots + \left(\frac{1}{2}\right)^{23} \times b_{23} \tag{6.2}$$

The functions developed to perform the conversion from the *unsigned long integer* representation of IEEE SP to the *double* format of the compiler and vice-versa are listed in Appendix B.

6.3 System Simulation Overview

The FT mode of the real-time DSP system architecture developed consists of TMR linear arrays consisting of identical processing elements, where each array is driven by an individual clock globally synchronized via a fault-tolerant clock system. In order to achieve tolerance of single permanent or transient faults during real-time DSP operations, static hardware redundancy is featured throughout the system in the form of TMR. The non-FT mode of this architecture consists of ignoring the TMR features and forming these processing elements into a single superarray for higher performance.

Each processing element of this dual-mode architecture consists of a DSP96002 microprocessor with its own local internal (and optional external) memory, as shown in Figure 6.2.

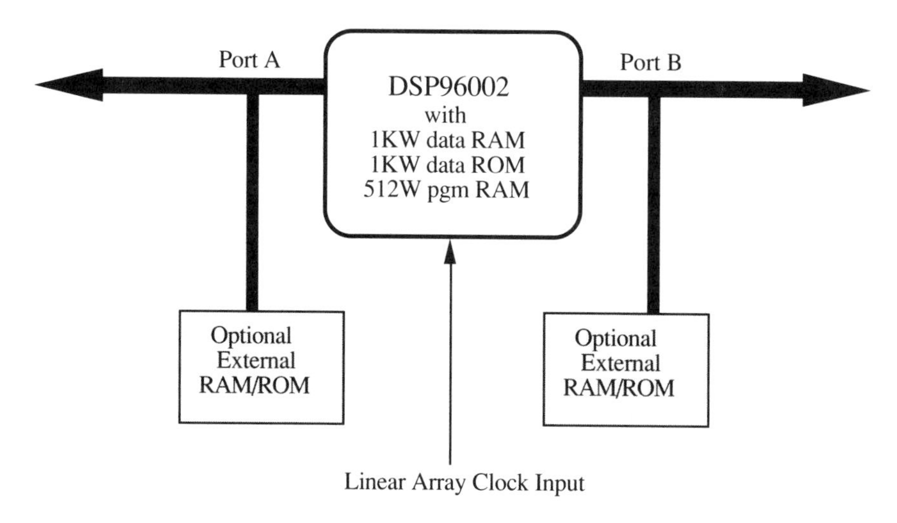

FIGURE 6.2 Organization of Each PE

Based on resource limitations associated with the host computing system for this simulation, the number of processors used is limited to ten or less. However, this in no way implies that the system itself is limited in this fashion. In fact, based on design choices made thus far, the system exhibits a high level of scalability approaching a linear scale. Thus, while a simulation may use only nine processors, the results for an equivalent system with ninety or even nine-hundred processors can be deduced.

The execution of multiple processors using a common external reference clock can be simulated using code based upon the following fragment:

```
while ( condition )                          /* loop while condition holds      */
  for (i=0; i<NUM_PE; i++) fsp_exec(i);      /* exec one cycle per processor    */
```

There are three significant simulation issues in addition to those associated with multiprocessor simulation. These are the simulation of: PE-to-PE communication; PE-to-output communication and voting; and input-to-PE communication.

6.4 Simulation of PE-to-PE Communication

In order to simulate the communication previously described between processors in this dual-mode system, the primary method is one of simulating individual hardware pins. That is, based on the PE-to-PE interface design developed, the simulation software will generally mimic these hardware interconnections.

Constant value connections to pins are simulated by executing a *fsp_wpin* function call for each relevant processor during every clock cycle using this value as the data parameter (see Appendix A for simulator function descriptions). Pin to pin connections are handled by executing a *fsp_rpin* and a *fsp_wpin* function call on the source and destination processors respectively during every clock cycle. Port to port connections are similarly handled using the *fsp_rport* and *fsp_wport* functions. Finally, the address decoder is simulated by using conditional statements which assert the destination pin if and only if a proper address is on the address bus.

There is one exception to the hardware mimic method described above that is used in this simulation. Due to the fact that the feedback status information read by the master processor is memory-mapped to a memory location on that master, the simulation library fails to correctly handle the port B data received during the *ICS register read* HI operation. Instead, it uses whatever value happens to be in that local memory location prior to the operation. In order to overcome this limitation, the simulation software developed performs a *fsp_rreg* function call to retrieve the ICS value from the slave, and then uses a *fsp_wmem* function call to write this status value directly to the local memory location on the master processor. This provides a simpler solution than modifying the simulation memory management software to overcome this limitation, while still accurately simulating the behavior of the hardware.

The actual function developed to perform PE-to-PE communication is included in the C simulation program developed for this project. This program is listed in Appendix B.

6.5 Simulation of PE-to-Output Communication and Voting

The simulation software for PE-to-output communication consists of two basic components. These are simulation of the transmission to the data latch and the voting itself. Each of these occurs during the execution of every clock cycle.

Since the Y:SYS_OUT location on the PE is memory-mapped to the data latch, the software can be made to simulate this transfer by merely reading the Y:SYS_OUT memory location on the PE using the *fsp_rmem* function. That is, since memory-mapped access is the view from the PE, the latch need only be simulated using a single PE memory location.

The fault-tolerant hardware output voter is a simple but important component of this fault-tolerant real-time DSP system. Each fault-tolerant majority gate in the voter computes a single-bit output value based on a majority of three single-bit inputs. Since the data path between processors in this system is 32-bits wide, thirty-two or more of these gates may be used in parallel. Bitwise logical operations can be used to simulate output voting for both the data and strobe output values.

The actual code developed to perform the PE-to-output and voting operations is included in the C simulation program developed for this project. This program is listed in Appendix B.

6.6 Simulation of Input-to-PE Communication

A technique similar to that used for simulating PE-to-output communication will be used to simulate the input-to-PE communication in this system. Since the data latch used by the first stage processors to read their input is memory-mapped on these processors to location X:SYS_IN, the simulation can be simplified by merely executing the *fsp_wmem* function call using the system input value as the data parameter. This function is executed on each first stage processor following each assertion of their input ready strobes.

To simulate the input ready strobes, port A on the initial-stage processors is monitored. When the X:SYS_IN request appears, the appropriate strobe is then asserted. Since the simulation uses the single-input form of this system, only one of these initial-stage processors is actually monitored.

The actual code developed to perform these operations is included in the C simulation program developed for this project. This program is listed in Appendix B.

Now that the simulation software has been developed for this system, the last major step is the development and evaluation of application software for system testing. Many issues will be considered in this preliminary test and evaluation phase of the project, including communication primitives, data propagation tests, digital filtering tests, fast Fourier transformation tests, fault injection and analysis tests, and analytical analysis.

6.7 Summary

This chapter has presented and developed all of the mechanisms that are necessary to simulate the architecture of this new system. Based on these developments, a software prototype is constructed on which preliminary tests and evaluations can be based. This is accomplished without the prohibitive costs normally associated with a hardware prototype.

The first simulation issue was an overview of the Motorola simulation libraries and their support for systems of this type. This was followed by a brief discussion on floating-point conversion between the values stored in unsigned long integers by the simulator library functions and the values stored in IEEE floating-point format by the compiler used to develop the new simulation software.

The remainder of this chapter dealt with carrying out the simulation of this particular architecture. First, an overview of system simulation for multiple microprocessor systems was presented. This was followed by separate discussion on each of the key elements for simulating the hardware communication in this system, including PE-to-PE communication, PE-to-output communication with voting, and input-to-PE communication.

Taken together, this chapter has described all of the elements that are necessary to simulate the architecture of this new system. With the completion of the software developed based on these ideas, the next step is to begin the preliminary test and evaluation phase using this new software prototype. In order to develop and perform these tests and evaluations, a number of issues will be addressed, such as communication primitives, data propagation, digital filtering, fast Fourier transformations, fault injection and analysis, and analytical analysis.

Chapter 7
PRELIMINARY TEST AND EVALUATION

This chapter develops and describes the preliminary test and evaluation experiments carried out on this system using the software prototype developed in the previous chapter. In order to develop and perform these tests and evaluations, a number of issues are addressed, such as communication primitives, data propagation, digital filtering, and fast Fourier transformations. Finally, a discussion on reliability modeling is presented which provides a more analytical description of the fault-tolerant capabilities of this system.

7.1 Communication Primitives

Before proceeding with a discussion of system tests and evaluations, the software issues associated with communication in this system must first be addressed. As previously discussed, the communication primitives provided for this system will use blocking, unreliable, and single-element buffering. These techniques were chosen to make the optimum use of the communication support provided by the DSP96002 HI hardware.

In order to provide greater support for programming of this system, the primitives themselves will use the remote procedure call (RPC) model of message passing as outlined in section 3.3. However, instead of subroutines forming these primitives, they will instead be implemented by macros. Since the DSP96002 architecture itself is based on internal pipelined execution, and since subroutine calls tend to seriously degrade this technique, macros provide a higher performance solution, since they incorporate in-line code instead of subroutine branch and return. And, since these primitives take advantage of the DSP96002 HI hardware support, the macros themselves are very short in length, thereby minimizing the increased amount of memory associated with macro calls.

There are four macros that have been developed for system-level communication on this machine. These are *get_IN*, *get_PE*, *put_PE*, and *put_OUT*. Together, these macros support input-to-PE, PE-to-PE, and PE-to-output communication.

The *get_IN* macro is used by an initial-stage PE to perform input-to-PE communication. It uses a single parameter to indicate the DSP96002 register which will hold the input value upon completion of execution, and is called from assembly-language in the following manner:

```
        get_IN   d0.s          ; get system input and return result in d0.s
```

The source code for the *get_IN* macro is shown below:

```
get_IN   macro   REG           ; macro to perform get operation from
                               ; the system input (FT or o/w)
                               ; located previous in the linear array
                               ; e.g. called via 'get_IN D0.S'

         MOVE    X:SYS_IN,REG  ; move input to register via
                               ; memory read (treating the input as
                               ; a memory element) @ a special address
         endm
```

Either one or both of the *get_PE* and *put_PE* macros are used by every PE to perform their part of PE-to-PE communication. The *get_PE* macro performs input from a previous PE in the linear array, and uses a single parameter to indicate the register which will hold the input value upon completion. The *put_PE* macro performs output to the next PE in the linear array, and uses a single parameter to indicate the register from which the output will be sent. For example, a PE which needs to propagate data from its input to its output in a PE-to-PE fashion might issue the following macro calls:

```
        get_PE   d0.s          ; get input from previous PE and return result in d0.s
        put_PE   d0.s          ; send value in d0.s to the next PE in the linear array
```

The source code for both the *get_PE* and *put_PE* macros is shown below:

```
get_PE   macro   REG                 ; macro to perform get operation over
                                     ; host i/f port A from the previous PE
                                     ; in the linear array when it is ready
                                     ; e.g. called via 'get_PE D0.S'

_LP1     JCLR    #HRDF,X:HSRA,_LP1   ; loop while HRDF bit of the local
                                     ; HSR reg is zero on port A

         MOVEP   X:HRXA,REG          ; now that the data has been received
                                     ; by the host interface, go ahead
                                     ; and read it via the local HRX reg
                                     ; for port A
```

```
        endm

put_PE  macro   REG                     ; macro to perform put operation over
                                        ; host i/f port B to the next PE in
                                        ; the linear array when it is ready
                                        ; e.g. called via 'put_PE D0.S'

_LP1    JCLR    #TXDE,Y:R_ICS,_LP1      ; loop while TXDE bit of the remote
                                        ; ICS reg on the slave is zero, via
                                        ; host function 1000 for ICS read

        MOVEP   REG,Y:R_TX              ; now that the remote PE is ready to
                                        ; receive, go ahead and send via
                                        ; host function TX register write
                                        ; by writing to a local Y ext peripheral
                                        ; loc memory-mapped to TX on the slave
        endm
```

The *put_OUT* macro is used by a final-stage PE to perform PE-to-output communication. It uses a single parameter to indicate the register which will holds the output value to be sent, and is called from assembly-language in the following manner:

```
        put_OUT   d0.s          ; send value in d0.s to the system output
```

The source code for the *put_OUT* macro is shown below:

```
put_OUT macro   REG                     ; macro to perform put operation over
                                        ; host i/f port B to output latch
                                        ; e.g. called via 'put_OUT D0.S'

        MOVE    REG,Y:SYS_OUT           ; move register to output via
                                        ; memory write (treating the output as
                                        ; a memory element) @ a special address
        endm
```

In addition to these communication macros, other macros have been developed to support DSP applications such as digital filtering. These macros will be described as each preliminary test and evaluation is introduced in an upcoming discussion.

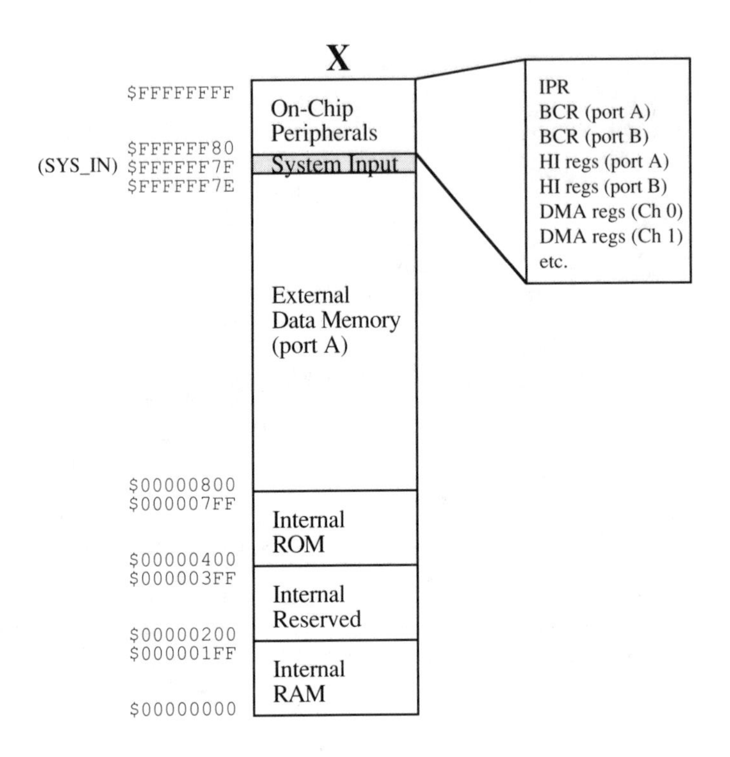

FIGURE 7.1 Memory map for the X address space

In addition to macros provided by the operating system, a number of equates have been defined for programming support. While many of these equates are predetermined by the DSP96002 architecture itself, some have been specifically developed for this system. Memory maps for both the X and Y data-memory address spaces are shown in Figures 7.1 and 7.2 respectively. The assembly-language equates themselves are listed in Appendix C.

With a general set of system macros and equates described to support input-to-PE, PE-to-PE, and PE-to-output communication, the next step in the development of this new system is that of preliminary tests and evaluations. These will include simple data propagation, digital filtering, and fast Fourier transformations.

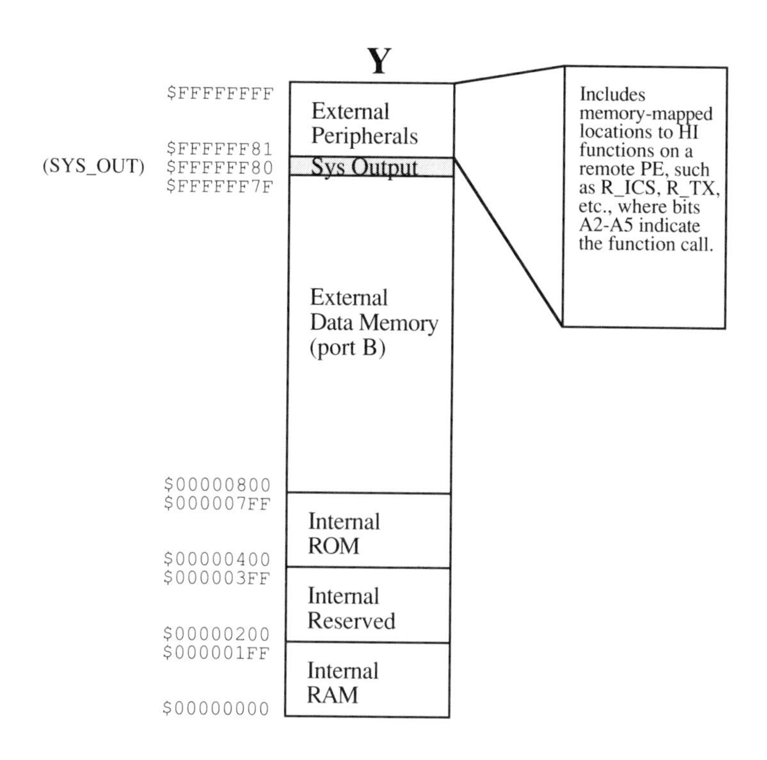

FIGURE 7.2 Memory map for the Y address space

7.2 Data Propagation

The first test chosen for this new system is a data propagation test. That is, the system is configured for operation, and the programs loaded on each processor in the system consist of reading a new input and immediately sending this input to their destination in the linear array. Exactly how the DSP96002 assembly-language test programs are structured to accomplish this propagation test depends upon the processor on which it will be executed.

Both dual-mode forms of system operation (i.e. FT or TMR mode, and non-FT or superarray mode) will be simulated for this test. In either case, and in general, there are three different forms for the software to be executed on all processors, depending upon whether the PE is an initial-stage, intermediate-stage, or final-stage processor.

In FT mode, there are three initial-stage processors, while in non-FT mode, there is only one. The data propagation test code executed by each initial-stage processor is shown below:

```
        title       "(PROP1.ASM) data propagation test for initial stage PEs"
        include     'OP_SYS.INC' ; include operating system equates & macros
        org         p:$100
START
        init_PE                         ; initialize this PE
LP      get_IN      d0.s                ; get system input
        put_PE      d0.s                ; transfer to next PE in linear array
        JMP         LP                  ; continue infinite loop
        STOP
```

The number of intermediate-stage processors depends upon the number of processors in the overall system. For example, in the 9-node system being simulated for these tests, there are three intermediate-stage processors in FT mode and seven in non-FT mode. The data propagation test code executed by each intermediate-stage processor is shown below:

```
        title       "(PROP2.ASM) data propagation test for intermediate stage PEs"
        include     'OP_SYS.INC' ; include operating system equates & macros
        org         p:$100
START
        init_PE                         ; initialize this PE
LP      get_PE      d0.s                ; get input from previous PE in linear array
        put_PE      d0.s                ; transfer to next PE in linear array
        JMP         LP                  ; continue infinite loop
        STOP
```

Finally, like the initial-stage processors, there are three final-stage processors in FT mode and only one in non-FT mode. The data propagation test code executed by each final-stage processor is shown below:

```
        title       "(PROP3.ASM) data propagation test for final stage PEs"
        include     'OP_SYS.INC' ; include operating system equates & macros
        org         p:$100
START
        init_PE                         ; initialize this PE
LP      get_PE      d0.s                ; get input from previous PE in linear array
        put_OUT     d0.s                ; transfer to the output latch
        JMP         LP                  ; continue infinite loop
        STOP
```

The results obtained from the data propagation tests indicate that twenty-four DSP96002 clock cycles are required for each new system output once the linear array has filled, for both the FT and non-FT modes of operation. This provides a basic description of the single-element I/O overhead associated with this system. Of course, as previously discussed, the throughput of the linear array system is primarily determined by the largest time requirement from any individual stage or PE. A brief summary of the raw data gathered from these tests is shown in Table 7.1.

TABLE 7.1 Summary of Data Propagation Results (9-node system)

Mode	Cycles to Fill Array	Cycles Between Outputs
FT mode	64	24
non-FT mode	208	24

7.3 Digital Filtering

The next test carried out on this new system is LTI IIR digital filtering. Since digital filtering is a fundamental operation for this fault-tolerant real-time DSP system, two macros have been developed to support basic DSP software development. Together, these macros implement one or more cascaded second-order difference equation sections in 1D form. The number of second-order sections, or *biquads*, to be executed on each processor is denoted by B.

The first macro is called *init_DF*, and its function is to initialize pointer and data registers. It includes three parameters: the number of biquads B; the address of the initialized delayed data values; and the address of the initialized filter coefficients. The second macro is called *calc_DF*, and its function is to perform the calculations and updates for one input and output of the digital filter. It includes two parameters: the number of biquads B; and the register to be used for input and output data. The DSP96002 assembly-language code for both macros is listed in Appendix C.

A total of eight IIR digital filtering test programs were carried out on the simulated system. These consisted of B = 1, 2, 3, and 4 in both FT and non-FT modes. As with the data propagation tests, the programs themselves varied depending on the stage of the processor on which the code will execute. Other than these communication macro calls, all processors execute the same software. The complete source code for these programs is provided in Appendix D.

For example, the program to implement a 2nd-order digital filter on a given processor in the FT or non-FT linear array requires two delayed data elements and four coefficients. On a 9-node system, this implies a 6th-order filter in FT mode and an 18th-order filter in non-FT mode.

Brief summaries of the raw data gathered from these tests for FT mode and non-FT mode are shown in Tables 7.2 and 7.3 respectively. In addition, the data in Table 7.4 represents these same tests on a single PE. A single PE was chosen as a means of comparison or frame of reference since this represents the most common comparable method currently used for digital filtering. This single-PE information was gathered on the simulated system by executing the digital filtering code on the initial-stage processors and the data propagation code on both the intermediate-stage and final-stage processors. The resulting cycles for filling the linear array were then decreased to negate the delay effect of all but the initial-stage processors.

TABLE 7.2 FT mode Digital Filtering Results (9-node system)

Biquads per Stage	Cycles to Fill Array	Cycles Between Outputs
B = 1 (6th-order)	132	42
B = 2 (12th-order)	152	50
B = 3 (18th-order)	172	58
B = 4 (24th-order)	204	66

TABLE 7.3 Non-FT mode Digital Filtering Results (9-node system)

Biquads per PE	Cycles to Fill Array	Cycles Between Outputs
B = 1 (18th-order)	384	42
B = 2 (36th-order)	440	50
B = 3 (54th-order)	496	58
B = 4 (72th-order)	600	66

TABLE 7.4 Single-PE Digital Filtering Results

Total Biquads	Cycles to Fill Array	Cycles Between Outputs
B = 3 (6th-order)	66	54
B = 4 (8th-order)	72	60
B = 6 (12th-order)	90	78
B = 8 (16th-order)	102	96
B = 9 (18th-order)	114	102
B = 12 (24th-order)	138	126
B = 16 (32nd-order)	168	156
B = 18 (36th-order)	186	174
B = 27 (54th-order)	258	246
B = 32 (64th-order)	294	288
B = 36 (72nd-order)	330	318

By using both the data propagation and IIR digital filtering data summary information, graphs can be generated which show the relationship between maximum sampling rate and filter order on this system. The filtering data is used to calculate the maximum sampling rate for varying non-zero filter orders, while the data propagation data describes a starting point in terms of a filter of order zero.

The significance of increasing the order of a digital filter lies in the precision and versatility of its processing capabilities. As the order of the filter is increased, the ability of the filter to select certain frequencies of the input signal to pass and certain ones to attenuate is improved. This improvement might be exhibited in terms of sharper transitions in the frequency domain from the pass band to the stop band, as well as less ripple in the spectrum itself. Thus, as the order of a filter is raised, so too is the potential capability of that filter to act upon its input signals in a more precise manner.

However, since the filter is implemented with a difference equation in the time domain, an increase in filter order implies an increase in processing requirements. Thus, the significance of larger maximum sampling rates for a digital filter implementation is that applications can be addressed with a wider bandwidth. Since the sampling rate must be at least twice the rate of the highest signal frequency to be filtered, the maximum sampling rate describes the upper bound of frequencies on which a particular digital system can operate. The larger the sampling rate, the wider the bandwidth of input signals that can be manipulated, and thus the more versatile the implementation.

The calculations for each of the maximum sampling rates is accomplished using the following equation:

$$f_s = \frac{1}{(no. of\ clk\ cycles\ per\ output) * (clk\ cycle\ period)} \tag{7.1}$$

For example, with a 9-node system made up of 40 MHz DSP96002 devices, the graph for both the FT and non-FT modes is shown in Figure 7.3 along with single-PE results for comparison. The plots of these three functions were obtained by substituting the "cycles between outputs" data from Tables 7.2-7.4 into equation 7.1 to compute the maximum sampling rate f_s.

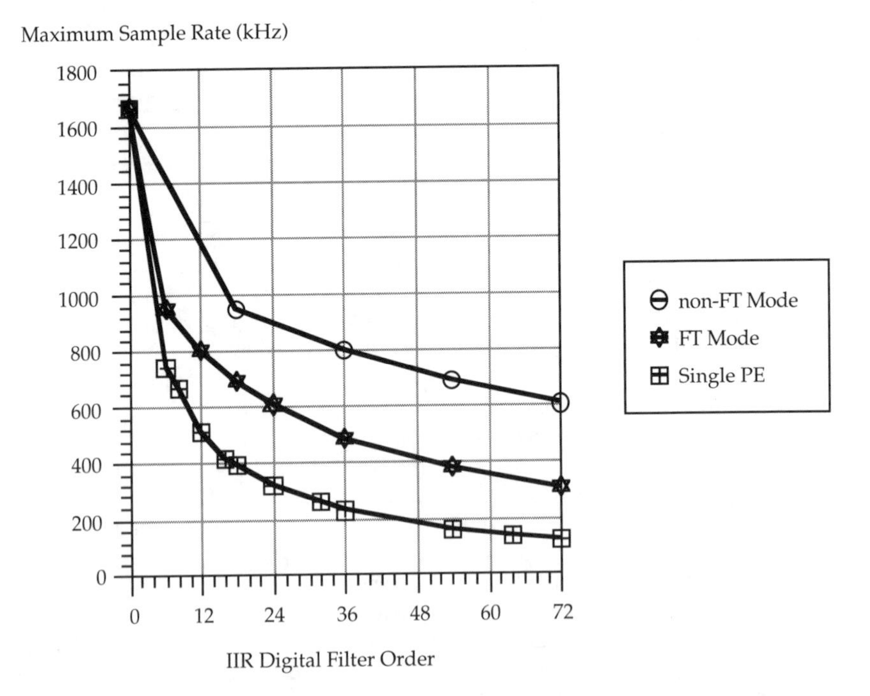

FIGURE 7.3 Maximum Sample Rate vs. Filter Order Performance Results for Both Modes of a 9-PE System

Although the filtering data tabulated for FT mode is based on tests extending to just 24th-order, this data is easily extrapolated to realize results for filters of a larger order (e.g. 72nd-order). This extrapolation is straightforward, since Table 7.2 indicates that each increase in filter complexity by one biquad per stage results in eight extra clock cycles being required to compute each new output.

These results indicate that a significant improvement in digital filtering sample rates has been achieved by both the FT and non-FT modes of this system, as compared with a traditional uniprocessor implementation. In both cases, these performance achievements make it possible for digital filters to be used with applications requiring sampling rates and involving frequencies beyond the capabilities of conventional uniprocessor systems.

7.4 Fast Fourier Transformations

The final performance test carried out on this new system is the fast Fourier transformation. Since the FFT is a fundamental operation for this fault-tolerant real-time DSP system, a general purpose macro has been developed to support basic DSP software development. This macro implements one or more stages or passes of a complex in-place radix-2 decimation-in-time FFT.

The macro is called *mpfft*, and was created by partitioning into parallel sections the uniprocessor code described in [DSP89]. It includes six parameters: the number of points N; the starting address of the data buffer; the starting address of the cosine and sine tables; the number of points in the cosine and sine tables; the starting pass; and the number of passes to be executed. In this way, the FFT itself can be partitioned between multiple processors, where each processor executes a subset of the total number of $\log_2 N$ passes required for the FFT. The code for this macro is listed in Appendix C.

A total of seven FFT test programs were carried out on the simulated system in FT mode. These consisted of N = 8, 16, 32, 64, 128, 256, and 512. In order to compare non-FT mode with FT mode, a 512-point FFT was carried out on the simulated system in non-FT mode. This value was chosen since it represents the smallest power of two divisible by nine (i.e. the number of processors in the superarray being simulated). In all FFT test cases, internal DSP96002 RAM is used for the programs (in program space), the real data (in X space), and the imaginary data (in Y space).

As with both the data propagation and digital filtering tests, the programs executing on the processors themselves varied depending on the stage of the processor on which the code will execute. Other than the communication macro calls and the starting pass specified for the FFT macro call, all processors execute essentially the same software. The complete source code for these programs is provided in Appendix D.

For example, the program to implement a 512-point FFT on a given processor in FT mode requires three passes on each processor for a total of nine passes. While the final-stage processors are calculating passes 7-9 on the first set of data, the intermediate-stage processors are calculating passes 4-6 on the second set of data and the initial-stage processors are calculating passes 1-3 on the third set of data.

A brief summary of the raw data gathered from the FT mode tests is shown in Table 7.5. By disabling all but the data transfer portions of the FFT routines, the data shown in Table 7.6 was determined (the 512-point test is intentionally omitted since, with double buffering, this test would exceed the limits of DSP96002 internal RAM). This data provides an estimate of the enhanced performance of the FFT tests when double-buffered DMA transfers are used (DMA functionality in the basic Motorola simulation object library is currently less than adequate). This estimate is reasonable since, as these results indicate, in all FFT test cases the data transfer time exceeds the calculation time, and thus the time required to transfer each new block of data points characterizes the overall performance of the FFT with double-buffered DMA.

The data gathered from the non-FT mode FFT test is shown in Table 7.7. Finally, the data in Table 7.8 represents all these same tests on a single PE. This single-PE information was gathered on the simulated system by executing the FFT code on the initial-stage processors and the data propagation code on both the intermediate-stage and final-stage processors. The resulting cycles for filling the linear array were then decreased to negate the delay effect of all but the initial-stage processors.

TABLE 7.5 FT mode FFT Results (9-node system)

Point Size	Cycles to Fill Array	Max. Cycles Between Each Output Set
$N = 8$	1178	692
$N = 16$	2150	1284
$N = 32$	4104	2370
$N = 64$	8112	4996
$N = 128$	16352	9848
$N = 256$	33282	19066
$N = 512$	68140	40222

TABLE 7.6 FT mode FFT Results (data propagation only)

Point Size	Cycles to Fill Array	Max. Cycles Between Each Output Set
$N = 8$	714	542
$N = 16$	1318	1054
$N = 32$	2538	2078
$N = 64$	4966	4126
$N = 128$	9834	8222
$N = 256$	19558	16414

TABLE 7.7 Non-FT mode FFT Results (9-node system)

Point Size	Cycles to Fill Array	Max. Cycles Between Each Output Set
$N = 512$	166958	37514

TABLE 7.8 Single-PE FFT Results

Point Size	Cycles to Fill Array	Max. Cycles Between Each Output Set
$N = 8$	516	858
$N = 16$	966	1692
$N = 32$	1902	3390
$N = 64$	3858	6882
$N = 128$	7986	14082
$N = 256$	16728	28968
$N = 512$	35196	59730

By using the results described in these tables, a graph can be generated which shows the relationship between maximum sampling rate and point size on this system as compared to a single processor implementation. As with digital filtering, the maximum sampling rate is determined from the number of clock cycles between each output set and the duration of each clock cycle.

The significance of increasing the point size of an FFT lies in the precision and versatility of its transformation capabilities. As the point size of the FFT is increased, the ability of the FFT to convert signals from the time domain to the frequency domain and back is improved. This improvement might be exhibited in terms of how well the frequency domain for a given time domain signal could be estimated or vice-versa. Thus, as the point size of an FFT is raised, so too is the potential capability of that FFT to transform its input signals in a more precise manner.

However, since the FFT is an algorithm whose complexity is directly related to the point size of the problem, an increase in point size implies an increase in processing requirements. Thus, like digital filtering, the significance of larger maximum sampling rates for an FFT implementation is that applications can be addressed with a wider bandwidth. Since the sampling rate must be at least twice the rate of the highest signal frequency to be transformed, the maximum sampling rate describes the upper bound of frequencies on which a particular digital system can operate. The larger the sampling rate, the wider the bandwidth of input signals that can be transformed to and from each domain, and thus the more versatile the FFT implementation.

The calculations for each of the maximum sampling rates is accomplished using the following equation:

$$f_s = \frac{1}{(no. of\ clk\ cycles\ per\ output\ set) * (clk\ cycle\ period) / (point\ size)} \tag{7.2}$$

For example, with a 9-node system made up of 40 MHz DSP96002 devices, the graph for FT mode (both non-DMA and estimated DMA) is shown in Figure 7.4 along with single-PE results for comparison. The plots of these three functions were obtained by substituting the "cycles between outputs" data from Tables 7.5, 7.6, and 7.8 into equation 7.2 to compute the maximum sampling rate f_S.

As with digital filtering, these results indicate that a significant improvement in FFT sample rates has been achieved by both the FT and non-FT modes of this system, as compared with a traditional uniprocessor implementation. In both cases, these performance achievements make it possible for FFT operations to be used with applications requiring sampling rates and involving frequencies beyond the capabilities of conventional uniprocessor systems.

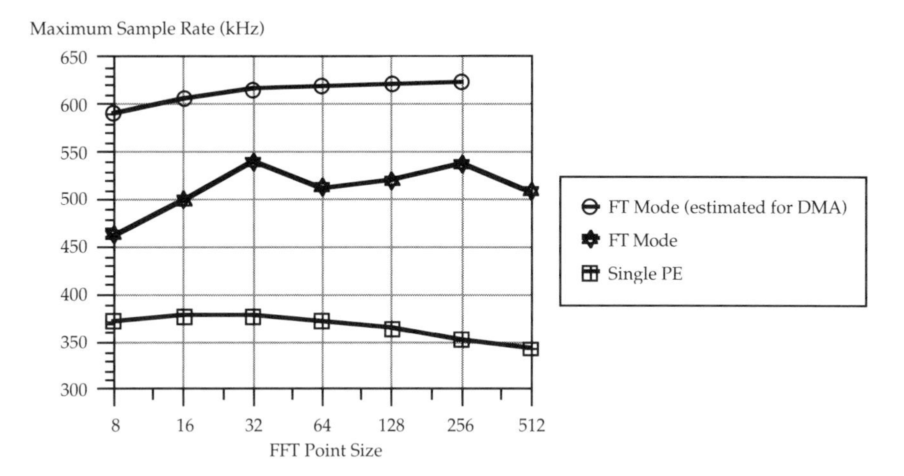

FIGURE 7.4 Maximum Sample Rate vs. Point Size Performance Results for FT Mode of a 9-PE System

The almost horizontal nature of these curves requires additional explanation. Unlike that with digital filtering, the FFT calculation algorithm used becomes more efficient as the point size increases, thereby tending to imply higher sampling rates for higher point sizes. However, as the point size increases, the data transfer or propagation requirements are also increased, which tends to reduce the sampling rate. These two factors together (computation versus I/O requirements) approximately balance, and thus the performance curves will tend to be reasonably flat for FFT test cases.

7.5 Fault Injection and Analysis

While the use of highly accurate multiprocessor simulation like that employed herein enables a system to undergo design, test, and evaluation at a minimal cost in terms of both man-hours and dollars, this simulation also provides a capability not normally associated with the real system. That is, one important by-product of this simulation method is the ability to inject one or more faults into the system and thereby analyze the system's behavior in their presence.

TABLE 7.9 Individual Fault Injection and Analysis Tests (FT mode)

Test	Description of How to Simulate
dead PE	disable that portion of the simulation code that establishes the connection between PEs; this consists of removing the port-to-port and pin-to-pin connection function calls.
insane PE	similar to the simulation of a dead PE, except that instead of disabling connections, the values on the connections are modified (e.g. by using a random number generator).
dead input-to-PE interface circuit	as with a dead PE, this is simulated by disabling the input connections of the appropriate PE.
insane input-to-PE interface circuit	as with an insane PE, this is simulated by modifying the values on the input connections of the appropriate PE.
dead PE-to-output interface circuit	as with a dead PE, this is simulated by disabling the output connections of the appropriate PE.
insane PE-to-output interface circuit	as with an insane PE, this is simulated by modifying the values on the output connections of the appropriate PE.
dead PE-to-PE interface circuit	as with a dead PE, this is simulated by disabling the input/output connections between the appropriate PEs.
insane PE-to-PE interface circuit	as with an insane PE, this is simulated by modifying the values on the input/output connections between the appropriate PEs.
dead linear array clock	disable that portion of the simulation code that executes the PEs in the appropriate linear array; this consists of removing PE execution function calls.
insane linear array clock	similar to the simulation of a dead linear array clock, except that instead of disabling PE executions, the array PEs are executed in a less than uniform manner with respect to the other two arrays (e.g. by using a random number generator).
transistor stuck-at-1 in FT voter	simulated separately from the system, using a logic simulator; this is preferable due to the simplicity of the voter circuit.
transistor stuck-at-0 in FT voter	same as that for stuck-at-1 faults.

For example, in a typical dynamic hardware redundancy design, many contingencies must be carefully tested to ensure that fault detection, location, and recovery are properly carried out. Often these types of designs are too complex to verify simply by using the system design itself. The ability to inject faults and analyze the system in the presence of faults thereby provides a major new solution to this difficult problem.

Although the fault-tolerant real-time DSP system under consideration is based on a static hardware redundancy model which employees fault masking as its primary fault-tolerance feature, fault injection and analysis can still be used to help verify correct fault masking in the system. While this technique is potentially more promising for other fault-tolerance models, it can nevertheless be useful.

A number of potential hardware faults can be considered for this system. A representative set of these possible faults has been identified for use in testing the behavior of the system. These fault tests are shown in Table 7.9, along with how their simulation can be carried out. The simulation of each of these individual fault injection and analysis tests has been carried out successfully on this system.

7.6 Analytical Analysis

While the previous test and evaluation developments have provided information on system performance and have helped to verify that the system is capable of tolerating single permanent or transient hardware faults, quantitative descriptions of system fault tolerance are desirable. These descriptions can be obtained by deriving analytical models for the system and comparing them to one or more alternatives, using techniques described in [JOHN89].

The failure rate for an electronic component or system is commonly described by the *bathtub curve* relationship. As shown in Figure 7.5, the failure rate decreases in the *early-life phase* of the system as substandard or inoperable components are identified. Similarly, the failure rate increases during the *wear-out phase* of the system as the useful life of the system is exceeded. However, by far the longest period of time is the *useful life period,* which is commonly approximated by a constant failure rate. It is this interval of time that is the most important, with the constant failure rate usually denoted by λ in units of failures per hour.

As previously discussed, the reliability $R(t)$ of a component or system represents the probability that the device will operate correctly throughout some interval $[t_0,t]$, given that the device was operating at initial time t_0. The assumption of a constant failure rate λ leads to the exponential failure law [JOHN89]:

$$R(t) = e^{-\lambda t} \tag{7.3}$$

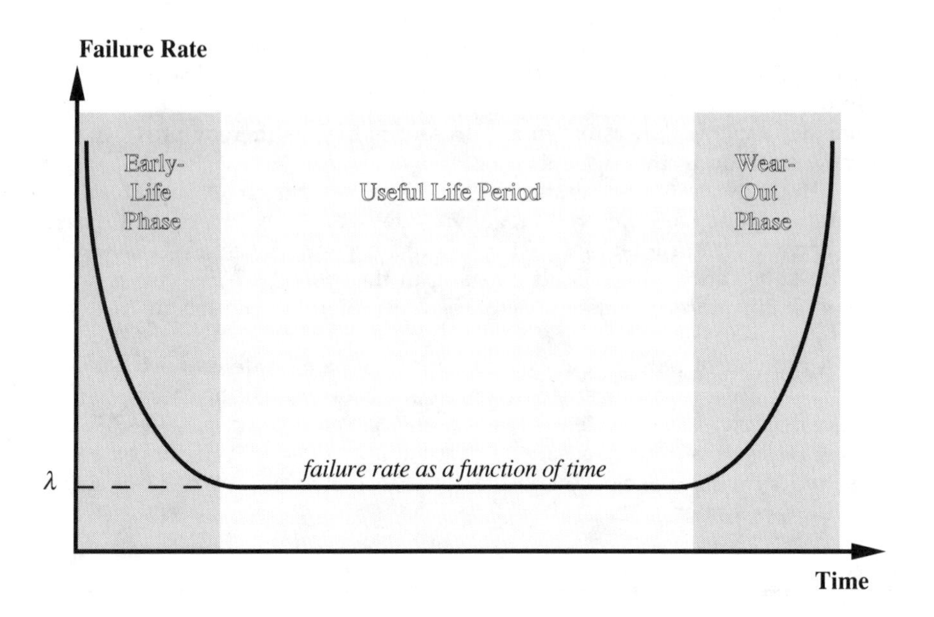

FIGURE 7.5 Bathtub Curve Relationship of Failure Rate vs. Time

The most widely used techniques for reliability analysis of systems are the analytical methods. Of these, the two most commonly used are the combinatorial and Markov modeling approaches. Combinatorial methods use probability theory to estimate the reliability of a system based on the reliability of its individual components and their interconnections. Markov models use the concepts of system state and state transition to develop a state diagram that describes the system.

Combinatorial models are more appropriate for systems which are constructed using series and parallel combinations of elements, while Markov models are superior for those complex systems which are extremely difficult to express in the form of probability theory. Since both the FT and non-FT modes of this system are relatively simple to describe in terms of series and parallel combinations of components, the combinatorial modeling method will be used.

The first step in the combinatorial modeling of this system for reliability analysis is the development of appropriate reliability block diagrams. These diagrams describe the series and parallel connections in the system. Series subsystems describe those components that form a subsystem chain, where the loss of any element in the chain will result in the failure of that entire subsystem. For example, the components in each FT mode linear array form a series subsystem, since the loss of any of these components will result in the loss of the entire array. By contrast, parallel subsystems describe those components in a subsystem which provide a level of redundancy, where the correct operation of any one of these components will allow the subsystem to operate.

The reliability of a series subsystem is equal to the product of the reliabilities of its components, and the failure rate is the sum of the component failure rates (assuming the exponential failure law holds). The reliability of a parallel subsystem is the product of the *unreliabilities* of its components (where the sum of the reliability and the unreliability of a component equals one). An important generalization of the parallel approach is the modeling of TMR systems. The reliability of a TMR subsystem made up of components with equal reliability $R(t)$ is [JOHN89]:

$$R_{TMR}(t) = R(t)R(t)R(t) + R(t)R(t)(1-R(t)) + R(t)(1-R(t))R(t) + (1-R(t))R(t)R(t)$$

$$\Rightarrow R_{TMR}(t) = 3R^2(t) - 2R^3(t) \quad (7.4)$$

In order to develop a reliability block diagram for each mode of this system, the following reliability variables will be used: $R_L(t)$ for one latch and associated circuitry; $R_{PE}(t)$ for one PE; $R_C(t)$ for one clock module; and $R_{FV}(t)$ for the fault-tolerant voter subsystem, where $R_T(t)$ indicates the reliability of each transistor in the fault-tolerant voter, and the voter consists of thirty-three individual TMR transistor circuits. The actual reliability of these components will vary depending on a number of factors, including maturity of component fabrication, amount of component screening, temperature, environment, number of pins, and the number of gates, transistors, or bits of memory [SIEW82].

The reliability block diagram for the FT mode of this system with k-stages is shown in Figure 7.6, while the reliability block diagram for the same system in non-FT mode is shown in Figure 7.7. The individual linear arrays in FT mode, and the superarray in non-FT mode, each consist of a single clock, two latches, and multiple PEs. The three linear array reliabilities in the FT mode are combined using the TMR reliability equation, and the voter reliability is described by a TMR equation raised to the 33rd power to account for the thirty-three independent single-bit fault-tolerant majority voter gates.

These reliability block diagrams lead to the following system reliability equations (one each for the FT and non-FT modes of operation):

$$R_{FT}(t) = \left(3R_C^2(t)R_L^4(t)R_{PE}^{2k}(t) - 2R_C^3(t)R_L^6(t)R_{PE}^{3k}(t)\right)\left(3R_T^2(t) - 2R_T^3(t)\right)^{33} \quad (7.5)$$

$$R_{nonFT}(t) = R_C(t)R_L^2(t)R_{PE}^{3k}(t) \quad (7.6)$$

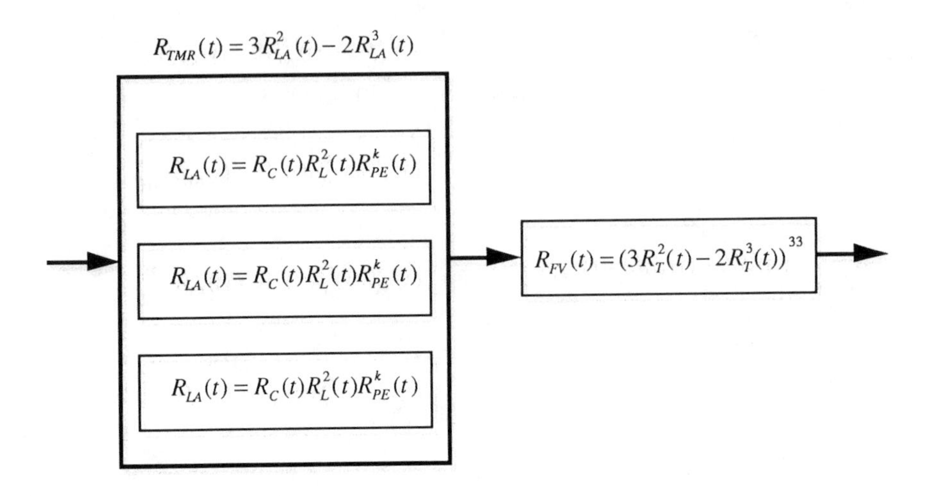

FIGURE 7.6 FT mode Reliability Block Diagram

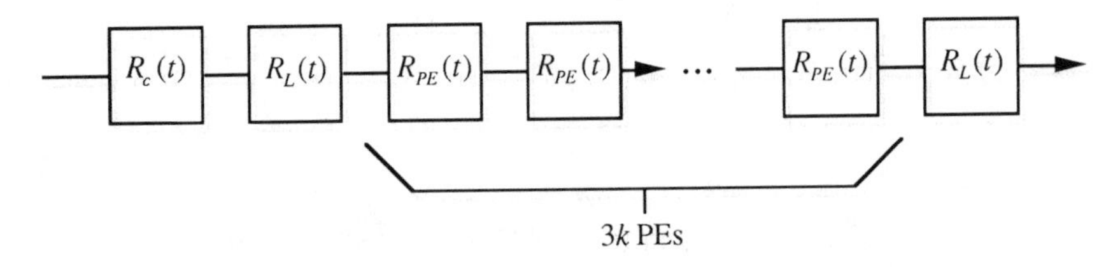

FIGURE 7.7 non-FT mode Reliability Block Diagram

An interesting comparison can be made between the FT mode of this system and a non-redundant linear array implementation comparable in performance. A comparable non-redundant system was chosen as a frame of reference in order to evaluate reliability factors while keeping performance factors equal. This non-redundant system is identical to the non-FT of this system with one-third the total number of processors. The reliability equation for the non-redundant implementation is shown below:

$$R_{nonred}(t) = R_C(t)R_L^2(t)R_{PE}^k(t) \tag{7.7}$$

In addition to reliability, there are many other evaluation parameters that can be considered. These include MTTF (mean time to failure), MTTR (mean time to repair), MTBF (mean time between failures), and MT (mission time). Of these, the last is the most appropriate for this fault-tolerant real-time DSP system. The mission time is the maximum amount of time of successful operation by the system before the reliability $R(t)$ falls below some predetermined level r.

In order to study the reliability and mission time capabilities of the system, the following example failure rates will be used to represent a *rough estimate* of average-quality components that might be used for hardware implementation:

$$\lambda_L = 0.3655 \times 10^{-6} \text{ failures per hour} \Rightarrow R_L(t) = e^{-0.3655\times10^{-6}t} \tag{7.8}$$

$$\lambda_{PE} = 50 \times 10^{-6} \text{ failures per hour} \Rightarrow R_{PE}(t) = e^{-50\times10^{-6}t} \tag{7.9}$$

$$\lambda_C = 0.1527 \times 10^{-6} \text{ failures per hour} \Rightarrow R_C(t) = e^{-0.1527\times10^{-6}t} \tag{7.10}$$

$$\lambda_T = 0.005 \times 10^{-6} \text{ failures per hour} \Rightarrow R_T(t) = e^{-0.005\times10^{-6}t} \tag{7.11}$$

Based on these component failure rate and reliability estimates, a comparison can be made between FT mode, non-FT mode, and a non-redundant counterpart to FT mode. Reliability graphs for both the first 1,000 and the first 10,000 hours of operation are shown in Figures 7.8 and 7.9 respectively.

These graphs clearly show the improvement in reliability provided by the FT mode of this system during the first 4500 hours of operation (i.e. approximately 6 months). After that point, the reliabilities of all three configurations become decreasingly significant. As the latter figure illustrates, it is common with TMR systems for the reliability of the redundant system to eventually drop below that of the non-redundant system. After some fixed amount of time, the redundant elements in the system eventually diminish in their effectiveness, and in fact eventually detract from overall system reliability as compared with the non-redundant system.

Consider the mission times of these three configurations at or above 99% reliability. The reliability equations indicate that the FT mode system meets this level for approximately 400 hours, while the non-redundant system only provides this level of reliability for approximately 60 hours. This represents an increase in mission time of approximately 340 hours, which is nearly a seven-fold improvement. A graph which shows the mission times of all three configurations at reliability levels exceeding 80% is shown in Figure 7.10.

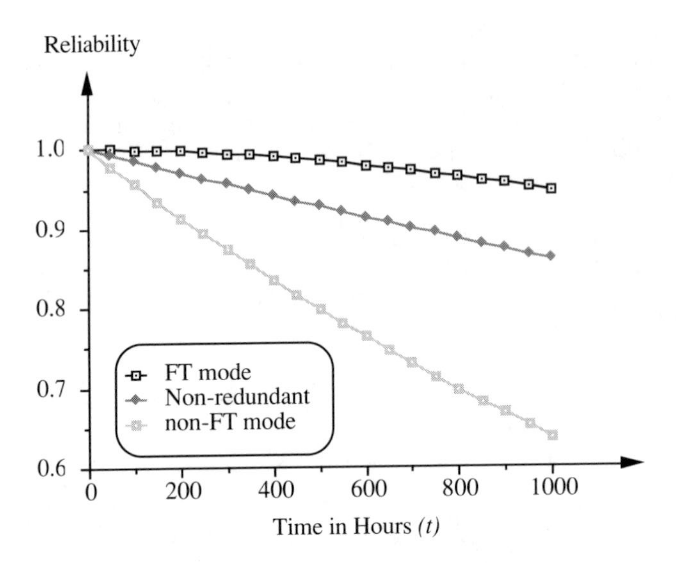

FIGURE 7.8 Reliability Comparison (first 1,000 hours of operation)

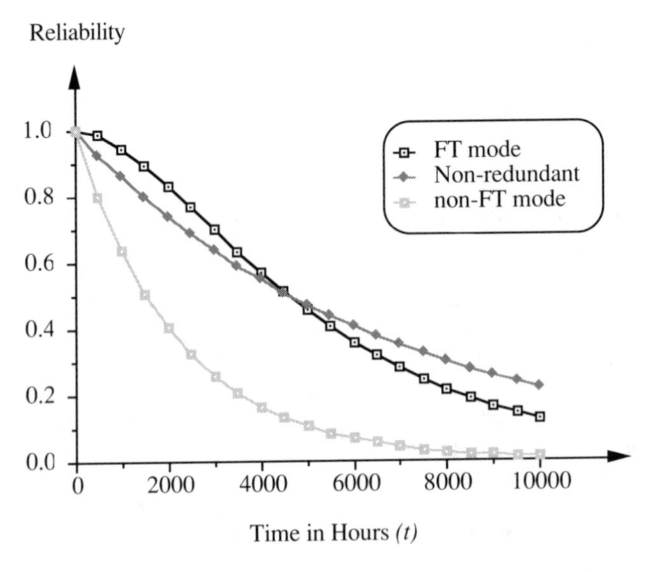

FIGURE 7.9 Reliability Comparison (first 10,000 hours of operation)

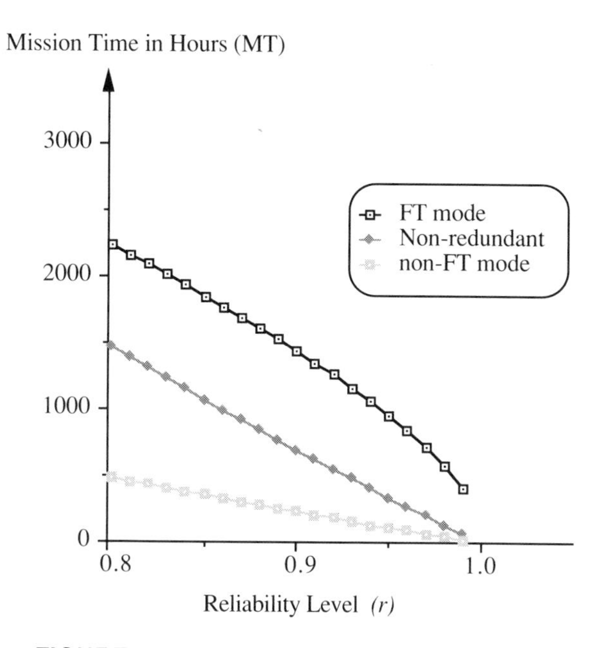

FIGURE 7.10 Mission Time Comparison

Thus, reliability modeling has shown a substantial improvement in reliability and mission time for the FT mode of this system versus a comparable non-redundant system. Based on relatively conservative failure rates that were used for different components in the system, improvements in mission time approaching an order of magnitude increase have been achieved. This represents a substantial improvement since, depending on the choice of components and their respective failure rates, mission times can be increased from days to weeks, from weeks to months, or even from months to years.

7.7 Summary

A number of major results have been described in this chapter. Communication primitives have been created for this system, and digital filtering and fast Fourier transformation macros have been partitioned and developed. These and other tests have been carried out on the system in a manner which accurately indicates actual system behavior. These tests have shown significant performance improvements, in terms of maximum sampling rates, over a comparable uniprocessor while still providing a high degree of growth potential.

Fault injection and analysis has been used to help verify the ability of this system to tolerate virtually any single permanent or transient hardware failure. Finally, analytical analysis has shown that the FT mode of this system provides a substantially larger level of reliability than a comparable non-redundant system, including mission time increases of nearly an order in magnitude.

In the final chapter that follows, a set of conclusions is presented which summarizes the important problems addressed by this text, the objectives that were identified, the contributions of this research, and future research directions.

Chapter 8
CONCLUSIONS

This text has addressed two important problems. The first is the need for DSP systems with higher levels of both performance and fault tolerance. The second is the need for methods to design and develop new multiprocessor and distributed systems in a way that reduces the escalating costs of this extremely complex task.

In addressing these two problems, three objectives were identified. The first objective was the creation of a new system for DSP which provides higher levels of performance and fault tolerance over traditional single-processor or non-redundant DSP designs. This objective was achieved by first considering the requirements of DSP applications, and then correlating them with fault tolerance and performance issues to produce the design for a new system, in terms of both hardware (architecture) and software (operating system communication primitives).

The second objective was the implementation of a new system prototype without the prohibitive costs normally associated with such designs. This objective was achieved by employing state-of-the-art simulation techniques to build a system prototype completely in software.

Finally, the third objective was to evaluate this new system with respect to both performance and fault tolerance. This final objective was achieved by using the system prototype developed in software, along with DSP applications such as digital filtering and fast Fourier transformations, to compare the new system with a uniprocessor system (for performance comparisons) and with a comparable non-redundant system (for reliability comparisons) using the same basic microprocessor.

The results of these tests have clarified many of the advantages offered by this new system. For example, using relatively simple digital filtering software partitioned and developed for this text, a significant improvement in sampling rate has been achieved versus a comparable DSP96002-based uniprocessor. Similar sampling rate performance increases have been achieved using relatively simple FFT software partitioned and developed for this text. In both cases, these performance achievements make it possible for DSP to be used with applications requiring sampling rates beyond the capabilities of conventional uniprocessors. In fact, the potential exists for even greater improvements by carefully customizing the software routines to optimally conform to this new architecture.

In addition to improvements in DSP performance, reliability modeling of this new system has shown a substantial improvement in reliability and mission time for this system versus a comparable non-redundant system. Based on relatively conservative failure rates that were used for different components in the system, improvements in mission time approaching an order of magnitude increase have been achieved. This represents a substantial improvement since, depending on the choice of components and their respective failure rates, mission times can be increased from days to weeks, from weeks to months, or even from months to years.

As exemplified in both performance and reliability, this system exhibits a high degree of growth potential. The architecture and other features of the system were designed so that additional processors can be added with little or no impact on other system features. For example, if this system with nine processors cannot quite satisfy the sampling rate requirements of a particular digital filtering application, it is likely that the same system with twelve processors can, without sacrificing the fault-tolerance features designed into the system.

Another important feature of this new system that is closely related to growth potential is flexibility. By incorporating a dual-mode approach, the system is capable of either fault-tolerant (FT) or traditional (non-FT or super-array) modes of operation. The choice of current system operation is based completely on the software executing on each processor. This makes it possible for the system to execute critical tasks in FT mode and less critical ones in non-FT mode, switching between modes at run time, with little or no overhead required for the transition.

The parallel architecture model used by this system is basically an MIMD processor array or grid architecture using a linear interconnection pattern between neighboring nodes. This system has the ability to support multiple instruction streams and multiple data streams flowing in either a unidirectional or bidirectional fashion. As is often the case with processor array architectures, a particular interconnection pattern was chosen for this system to exploit the concurrency of the algorithms to be used.

Of the two basic approaches to processor arrays (i.e. systolic arrays and wavefront arrays), the wavefront approach is perhaps the more suitable model since, like wavefront arrays, this system uses handshaking protocols to provide asynchronous interprocessor communication and operations. However, one important distinction between this system and the common methods of both systolic and wavefront arrays is the size and power of the individual nodes and the granularity of parallelism that is employed. While systolic and wavefront arrays are associated with fine-grained machines consisting of extremely small and simple PEs, this system instead uses a more medium-grained level of concurrency with advanced and extremely complex microprocessors serving as nodes.

This method could be described as starting with the best in DSP uniprocessors as the basis for designing a DSP multiprocessor. The use of other such advanced microprocessors has also been the basis of multiprocessor architectures like the Intel iPSC/1, iPSC/2, and iPSC/860, the Cal Tech Cosmic Cube systems, and the Thinking Machines CM-5. However, in contrast, all of these systems incorporate a hypercube interconnection structure.

This research has helped to remedy the two problems which inspired its initiation. That is, two major contributions or accomplishments have been identified. The first contribution is a new fault-tolerant computer system which is suitable for a wide variety of DSP applications, as well as other similar scientific computing problems. The architecture associated with this system is easily generalized, providing growth paths for both performance and reliability as application requirements dictate. These growth paths only tend to strengthen the long-term viability of this new system.

The second contribution is the concept, and its illustrated proof, that highly complex multiprocessor or distributed systems, both traditional and fault-tolerant, can be designed and developed at a relatively minimal cost. In effect, it has been shown that a prototype can be developed for a wide variety of such systems which is remarkably accurate in its demonstration of how the actual hardware-based system functions. This approach makes it possible for candidate multiprocessor computer systems to be constructed and evaluated without prematurely incurring the escalating costs of hardware implementation.

Finally, these two major contributions lead to two directions of future research that warrant attention. The first direction is one of further experimentation based on a hardware implementation of this new computer system. While the accomplishments of this research have postponed the need for hardware implementation, there will of course come a time when the actual system must be built.

The second direction of future research is quite distinct from that completed in this text. The research herein has essentially served as a proof of concept with respect to the potential of highly accurate software simulation. The next logical step is the study and development of a general-purpose, object-oriented computer environment for multiprocessor and distributed computer system design and experimentation. This new environment would allow the computer designer to actually identify and experiment with a variety of candidate systems, and eventually select the optimum candidate before any hardware implementations are necessary.

Appendix A
MULTIPROCESSOR SIMULATION LIBRARY FUNCTIONS

This appendix provides a brief description for each of the twenty-one C functions in the non-display portion of the Motorola DSP96002 microprocessor simulation object library. These functions are used as primitives by the multiprocessor simulation software developed for this text.

fsp_dasm is used to disassemble DSP96002 object code into assembly language. Parameters include opcodes and a return buffer.

fsp_exec is used to execute one clock cycle on a particular DSP96002 device, updating its device state data structure. Parameter consists of the DSP96002 device index.

fsp_free is used to free all memory and close all files allocated to a particular DSP96002 device. Parameter consists of the device index.

fsp_inall is used to initialize all DSP96002 device state data structures (via multiple calls to *fsp_init*). No explicit parameters.

fsp_init is used to initialize or reset the device state data structure of a particular DSP96002 device. Parameter consists of the device index.

fsp_ldall is used to load all DSP96002 device state data structures in a multiple device simulation from a set of state files. Parameter consists of a base name upon which the filenames are based.

fsp_ldmem is used to load the memory space of a particular DSP96002 device from an object file. Parameters consist of a device index and a filename.

fsp_load is used to load the device state data structure from a state file for a particular DSP96002 device. Parameters consist of a device index and a filename.

fsp_masm is used to assemble a DSP96002 assembly language instruction into object code. Parameters consist of a mnemonic, opcodes, and an error message.

fsp_new is used to create and initialize the device state data structure for a new DSP96002 device to be simulated. Parameters consist of a device index and a device type.

fsp_path is used to construct an alias name from a user-provided base name and suffix in order to simplify file access. Parameters consist of a base name, a suffix, and a new name.

fsp_rmem is used to read the contents of a selected memory location in a particular DSP96002 device. Parameters consist of a device index, a memory designator, an address, and a return value.

fsp_rpin is used to read the value of a selected pin on a particular DSP96002 device. Parameters consist of a device index, a pin designator, and the pin value.

fsp_rport is used to read the state from a selected port on a particular DSP96002 device. Parameters include a device index, a port number, a return data value, and a return force value which indicates which port bits are actually being driven by the device as outputs.

fsp_rreg is used to read the contents of a selected register in a particular DSP96002 device. Parameters include a device index, a register number, and a return value.

fsp_save is used to save the device state data structure for a particular DSP96002 device to a file. Parameters include a device index and a filename.

fsp_svall is used to save the device state data structures of all DSP96002 devices to a set of files. Parameter consists of a base filename.

fsp_wmem is used to write a value to a selected memory location in a particular DSP96002 device. Parameters include a device index, a memory designator, an address, and the value to write.

fsp_wpin is used to write a value to a selected pin on a particular DSP96002 device from outside that device. Parameters include a device index, a pin number, and the value to write.

fsp_wport is used to write or force data onto a selected port on a particular DSP96002 device from outside that device. Parameters include a device index, a port number, the port data to write, and the port forcing state to indicate which port bits are actually to be driven as inputs.

fsp_wreg is used to write a new value into a selected register in a particular DSP96002 device. Parameters include a device index, a register number, and the value to write.

Appendix B
SIMULATION SOFTWARE LISTINGS

This appendix provides a complete listing of the C multiprocessor simulation software developed for this text. This software serves as the system prototype upon which tests and evaluations have been undertaken.

```
/* (SYS_SIM.C) FTDSP Simulation Source Code                                */
/*                                                                         */
/*      This file simulates the operation of a dual-mode system.           */
/*      The first mode is an FT mode, which consists of a TMR              */
/*      linear array system with 3-stage arrays.  Each PE consists         */
/*      of a DSP96002 and local memory (internal and optional external),   */
/*      and each array has its own clock circuit.  Hardware voting         */
/*      is provided at the system output, where a FT voter is              */
/*      used.  The second mode uses these same PEs to implement a 9        */
/*      processor "superarray" for higher performance.  See below          */
/*      for a sketch of each mode's configuration.                         */
/*                                                                         */
/*                                                                         */
/* FT MODE OF OPERATION:                                                   */
/*                                                                         */
/*              +-----> PE0 -----> PE3 -----> PE6 ---+                     */
/*              |                                     \                    */
/*  sys_input ---+-----> PE1 -----> PE4 -----> PE7 -----> V -> FT_out      */
/*              |                                     /                    */
/*              +-----> PE2 -----> PE5 -----> PE8 ---+                     */
/*                                                                         */
/*                                                                         */
/* non-FT MODE OF OPERATION:                                               */
/*                                                                         */
/* sys_input ---+-----> PE0 -----> PE3 -----> PE6 -------+                 */
/*                                                       |                 */
/*              +<---------------------------------------+                 */
/*              |                                                          */
/*              +-----> PE1 -----> PE4 -----> PE7 ------+                  */
/*                                                      |                  */
/*              +<--------------------------------------+                  */
/*              |                                                          */
/*              +-----> PE2 -----> PE5 -----> PE8 -----> nonFT_output      */
/*                                                                         */
/***************************************************************************/
/* EDIT HISTORY                                                            */
/* ----------------------------------------------------------------------- */
/* 01/11/91 adg created SYS_SIM.C                                          */
/* 01/15/91 adg modified to reflect use of my ATP approach                 */
/* 01/15/91 adg modified to reflect use of voting output synch. using      */
/*              the force parm. of the fsp_rport function                  */
/* 01/17/91 adg modified to effect operation of simple test                */
/* 01/27/91 adg tests to determine if a value sent to a Y external I/O     */
/*              address appears on the port B address & data buses (yes)   */
/*              also tried a random X external memory address (yes)        */
/*              also added tmp code to display port data/addr info         */
/* 01/27/91 adg major mods to PE_to_PE to begin to sim the i/f design      */
/* 01/28/91 adg changed sys I/O to use of X:SYS_IN & Y:SYS_OUT mem refs */
/*              debug of in->PE, PE->PE, and PE->voter code parts          */
```

```
/*              (with good results so far)                                  */
/* 01/29/91 adg assorted repairs                                            */
/* 01/31/91 adg added code to require and provide acknowledgements for      */
/*              final stage outputs; this keeps each final stage PE in      */
/*              each array from continuing until at least two out           */
/*              values that agree are obtained.  This is provided to        */
/*              solve the problem of the ATP arrays getting so far out      */
/*              of synch. that they impose on the TMR voting.               */
/* 02/04/91 adg minor modifications                                         */
/* 02/06/91 adg modified PE_to_PE to use a more effective means for         */
/*              overriding the way the sim. library routines want to        */
/*              treat all memory-mapped locations as memory...in            */
/*              particular, to make the R_ICS remote ICS read work.         */
/*              This is done by fsp_rreg of ICS on dst and fsp_wmem         */
/*              write to R_ICS location on src processor.                   */
/* 02/12/91 adg Minor repairs.                                              */
/* 02/13/91 adg Beginnings of adding dual-mode simulation support.          */
/* 02/14/91 adg First test of non-FT mode execution.  Simple data           */
/*              propagation as used w/ the first FT mode test.  Code        */
/*              modified so that the first PE (PE0), middle PEs             */
/*              (PE1-PE7), and the last      PE (PE8) each run different    */
/*              code, like that of the 3 stages of the FT mode test.        */
/* 02/18/91 adg Added support for the Input_Ready strobe.  This is          */
/*              used to synchronize the input latch(es) between the         */
/*              system input(s) and the 1st (stage) PE(s).                  */
/* 02/18/91 adg Also added support for the nonFT_Output_Ready strobe.       */
/* 02/23/91 adg Disabled clock divergence for testing.                      */
/* 02/26/91 adg Modifications to support synch instead of asynch clocks.    */
/* 02/27/91 adg Also modified for both output strobes.                      */
/* 03/10/91 adg Added connection of slave HR* to master IRQA* for dma       */
/*              access during PE->PE transfers                              */
/* 03/23/91 adg Modified output statements for more fp precision.           */
/* 03/23/91 adg Modified data input sim to read data from a disk file.      */
/*                                                                          */
/************************************************************************** */

#include "fconfig.h"            /* simulator include files */
#include "fbracket.h"
#include "fspexec.h"
#include "fspsdef.h"
#include "fspsdcl.h"
#include "fsppin.h"
#include "stdlib.h"             /* C include files and O/S support */
#include "stdio.h"
#include "time.h"

#define NUM_PIPE    3                          /* Number of Processor Arrays */
#define NUM_STAGE   3                          /* Number of Stages per Array */
#define NUM_PE      NUM_PIPE*NUM_STAGE         /* No. of DSP96002 processors */
```

```
#define SYS_IN      0xFFFFFF7FL     /* X addr for PE input */
#define SYS_OUT     0xFFFFFF80L     /* Y addr for PE output */
#define R_ICS       0xFFFFFFE0L     /* Y addr for remote ICS access (r/w) */
#define R_TX        0xFFFFFFE8L     /* Y addr for remote TX reg transfer */

#define IN_USE      0xFFFFFFFFL     /* Port force constant => in use */
#define IDLE        0x00000000L     /* Port force constant => not in use */
#define TRUE        1               /* True constant for I/O strobes */
#define FALSE       0               /* False constant for I/O strobes */
#define DELAY       2               /* Used for delayed strobe T values */

main()                          /* start of main program */
{
  int i,j,err,ts,s1,s0,out[NUM_PE];
  int Input_Ready=FALSE, FT_Output_Ready=FALSE, nonFT_Output_Ready=FALSE;
  long int cycles;
  struct fsp_var *fspn[NUM_PE]; /* dsp device pointers */
  FILE *in_file,*fopen();

  unsigned long in1, in2, in3, FT_out=0, pforce1, pforce2, pforce3;
  unsigned long cnvt_to_IEEE(), ackval=0xFFFFFFFFL, sys_input, nonFT_out;
  unsigned long addr,frc;
  char sav_state[3];
  double cnvt_to_double(),dnum;
  unsigned long tmp1,tmp2;
  char tmp3;

  int mem_none;
  unsigned long val0,val1,val2;

  fsp_const.num_fsps=NUM_PE;          /* setup simulation s/w variables */
  fsp_const.fsp_list= fspn;

  printf("How many cycles to execute? ");
  scanf("%ld",&cycles);
  printf("Save processor state files upon completion [Y/N] ? ");
  scanf("%s",sav_state);
  printf("Using %d processors for %ld cycles.\n",NUM_PE,cycles);
  srand(clock());          /* set a seed for the RNG rand() func via clock */

  in_file=fopen("sysin.dat","r");          /* open file for data input to system */
  fscanf(in_file,"%lf",&dnum);             /* get the first value */
  sys_input=cnvt_to_IEEE(dnum);            /* convert and store in IEEE fmt */

  /* allocate memory for each device */

  printf("Allocating memory for each processor...\n");
  for (i=0;i<NUM_PE;i++) fsp_new(i,DSP96002);
```

```
 /* load device program for each device */

 printf("Loading object code for each processor to execute...\n");

 /* FT mode TEST LOADS */
 for (i=0;i<3;i++) fsp_ldmem(i,"st1.lod");        /* FT mode stage 1 PEs */
 for (i=3;i<6;i++) fsp_ldmem(i,"st2.lod");        /* FT mode stage 2 PEs */
 for (i=6;i<9;i++) fsp_ldmem(i,"st3.lod");        /* FT mode stage 3 PEs */

 /* non-FT mode TEST LOADS */
/*  fsp_ldmem(0,"st1.lod"); /* load object code for non-FT input PE */
/*  for (i=1;i<8;i++) fsp_ldmem(i,"st2.lod"); /* load code for middle PEs */
/*  fsp_ldmem(8,"st3.lod"); /* load object code for non-FT output PE */

/*  fsp_ldmem(0,"st1.lod"); /* load object code for non-FT input PE */
/*  fsp_ldmem(1,"st2.lod"); /* load object code for non-FT intermediate PE */
/*  fsp_ldmem(2,"st3.lod"); /* load object code for non-FT intermediate PE */
/*  fsp_ldmem(3,"st4.lod"); /* load object code for non-FT intermediate PE */
/*  fsp_ldmem(4,"st5.lod"); /* load object code for non-FT intermediate PE */
/*  fsp_ldmem(5,"st6.lod"); /* load object code for non-FT intermediate PE */
/*  fsp_ldmem(6,"st7.lod"); /* load object code for non-FT intermediate PE */
/*  fsp_ldmem(7,"st8.lod"); /* load object code for non-FT intermediate PE */
/*  fsp_ldmem(8,"st9.lod"); /* load object code for non-FT output PE */

 /* Change BCR (bus cntl reg) value for each device to 0 */

 printf("Changing the BCR value in each processor to zero...\n");
 for (i=0;i<NUM_PE;i++) fspn[i]->fsp_reg[MR_ABCR]=
                 fspn[i]->fsp_reg[MR_BBCR]=0L;

 /*************************************************************************** */
 /* Loop for the prescribed number of clock cycles.  For each clock          */
 /* cycle, each processor executes a cycle, where a RNG function             */
 /* is used to simulate the clock divergence problem.  Also, in->PE,         */
 /* PE->output, and interprocessor communication is provided.                */
 /* And, of course, the system output voter (simulated by a single           */
 /* voter) is simulated.                                                     */
 /*************************************************************************** */

 printf("Executing %ld clock cycles in parallel on all processors:\n",cycles);

 while (fspn[0]->fsp_reg[REG_CYCL]<cycles) {        /* loop until cycles used    */

  for (i=0;i<NUM_PE;i++) {              /* simulate constant pin connections */
    fsp_wpin(i,PIN_ata,PINVAL_L);      /* assert Port A TA* ; no wait states */
    fsp_wpin(i,PIN_bta,PINVAL_L);      /* assert Port B TA* ; no wait states */
  }
```

```
if (Input_Ready==TRUE) {                        /* If Input Latch free for new data */
 for (i=0;i<NUM_PIPE;i++)                       /* Transfer System to Input Latches */
   fsp_wmem(i,MEM_X,SYS_IN,&sys_input);              /* simulated via mem write */
 Input_Ready=FALSE;                             /* & show latch is now not free. */
 fscanf(in_file,"%lf",&dnum);                   /* Get a new input value */
 sys_input=cnvt_to_IEEE(dnum);                  /* convert and store in IEEE fmt */
}

for (i=0;i<NUM_PE;i++)                          /* loop for all PEs */
   fsp_exec(i);                                 /* execute one clock cycle */

PE_to_PE(0,3);      /* interprocessor xfer from PE0 to PE3 */
PE_to_PE(1,4);      /* interprocessor xfer from PE1 to PE4 */
PE_to_PE(2,5);      /* interprocessor xfer from PE2 to PE5 */
PE_to_PE(3,6);      /* interprocessor xfer from PE3 to PE6 */
PE_to_PE(4,7);      /* interprocessor xfer from PE4 to PE7 */
PE_to_PE(5,8);      /* interprocessor xfer from PE5 to PE8 */
PE_to_PE(6,1);      /* interprocessor xfer from PE6 to PE1 (non-FT mode) */
PE_to_PE(7,2);      /* interprocessor xfer from PE7 to PE2 (non-FT mode) */

/*-------------------------------------------------------------------------*/
/* TEMPORARY FOR VIEWING PE PORT OUTPUTS */
/*    printf("CLOCK CYCLE %ld\n",fspn[0]->fsp_reg[REG_CYCL]);
  printf("PE  A.addr  force  A.data  force  B.addr  force  B.data   force\n");
  printf("------------------------------------------------------------------\n");
  for (j=0;j<NUM_PE;j++) {
     fsp_rport(j,PORT_AA,&tmp1,&tmp2);
    printf("%2d %8lx %8lx ",j,tmp1,tmp2);
     fsp_rport(j,PORT_AD,&tmp1,&tmp2);
    printf("%8lx %8lx ",tmp1,tmp2);
     fsp_rport(j,PORT_BA,&tmp1,&tmp2);
    printf("%8lx %8lx ",tmp1,tmp2);
     fsp_rport(j,PORT_BD,&tmp1,&tmp2);
     printf("%8lx %8lx\n",tmp1,tmp2);
  }
  printf("------------------------------------------------------------------\n");
     printf("PE0\tPE1\tPE2\tPE3\tPE4\tPE5\tPE6\tPE7\tPE8\n");
  for (j=0;j<NUM_PE;j++) {
    fsp_rreg(j,REG_PC,&tmp1);
     printf("%lx\t",tmp1);
  }
   printf("\n");
   scanf("%c",&tmp3);
/*-------------------------------------------------------------------------*/

  fsp_rmem(6,MEM_Y,SYS_OUT,&in1);        /* xfer from PE6 to in1 for voter */
  fsp_rmem(7,MEM_Y,SYS_OUT,&in2);        /* xfer from PE7 to in2 for voter */
  fsp_rmem(8,MEM_Y,SYS_OUT,&in3);        /* xfer from PE8 to in3 for voter */
```

```
  nonFT_out=in3;                       /* non-FT mode output right from PE8 */

  FT_out=(in1&in2)|(in1&in3)|(in2&in3);         /* perform the voting */

                                                /* HANDLE Input_Ready STROBE SIM: */
   fsp_rport(0,PORT_AA,&addr,&frc);             /* read port A addr bus on stage 1 PE */
  ts=fsp_rpin(0,PIN_ats);                       /* read TS* port A on stage 1 PE */
  s1=fsp_rpin(0,PIN_as1);                       /* read S1 port A on stage 1 PE */
  s0=fsp_rpin(0,PIN_as0);                       /* read S0 port A on stage 1 PE */
                                                /* where: S1=0, S0=1 => X mem */
  if (ts==PINVAL_L && addr==SYS_IN && frc==IN_USE
            && s1==PINVAL_L && s0==PINVAL_H)
                                                /* if access to X:SYS_IN happening, */
   Input_Ready=TRUE;                            /* this means input latch being read */
                                                /* and thus available for new data */


                                       /* HANDLE BOTH Output_Ready STROBES */
  for (i=NUM_PE-3;i<NUM_PE;i++) {      /* loop for each final-stage PE */
    if (out[i]==DELAY)                 /* First make use of tmp delay value */
     out[i]=TRUE;                      /* so true is delayed one clk cycle */
   else out[i]=FALSE;                  /* o/w new output not ready */
     fsp_rport(i,PORT_BA,&addr,&frc);           /* read port B addr bus on PE */
    ts=fsp_rpin(i,PIN_bts);                     /* read TS* port B on PE */
   s1=fsp_rpin(i,PIN_bs1);                      /* read S1 port B on PE */
   s0=fsp_rpin(i,PIN_bs0);                      /* read S0 port B on PE */
                                                /* where: S1=S0=0 => Y mem */
    if (ts==PINVAL_L && addr==SYS_OUT && frc==IN_USE
            && s1==PINVAL_L && s0==PINVAL_L)
                                                /* if access to Y:SYS_OUT happening, */
     out[i]=DELAY;                              /* means output latch being written */
                                                /* and thus available new output */
                                                /* data after a cycle delay */
  }
  if (out[NUM_PE-1]==TRUE)                      /* if last PE has asserted out strobe */
    nonFT_Output_Ready=TRUE;                    /* set non-FT mode output strobe to T */
  else nonFT_Output_Ready=FALSE;                /* else set F */
  if ( (out[NUM_PE-3]==TRUE && out[NUM_PE-2]==TRUE)||   /* vote on FT mode */
      (out[NUM_PE-3]==TRUE && out[NUM_PE-1]==TRUE)||    /* output strobe */
      (out[NUM_PE-2]==TRUE && out[NUM_PE-1]==TRUE) )    /* & if majority T */
    FT_Output_Ready=TRUE;                               /* then set strobe T */
  else FT_Output_Ready=FALSE;                           /* else set F */


/*
   printf("CLK:%5ld ",fspn[0]->fsp_reg[REG_CYCL]);
   printf("INPUT:%+13e  ",cnvt_to_double(sys_input));
   printf("FT_OUT:%+13e  ",cnvt_to_double(FT_out));
   printf("nonFT_OUT:%+13e\n",cnvt_to_double(nonFT_out));
```

```
    if (Input_Ready==TRUE) printf("FT & non-FT Input Now Ready.\n");
    if (FT_Output_Ready==TRUE) printf("FT Output Now Ready.\n");
    if (nonFT_Output_Ready==TRUE) printf("non-FT Output Now Ready.\n");
*/

    if (nonFT_Output_Ready==TRUE || FT_Output_Ready==TRUE) {
      printf("CLK:%5ld ",fspn[0]->fsp_reg[REG_CYCL]);
      printf("INPUT:%+13e  ",cnvt_to_double(sys_input));
    }
    if (FT_Output_Ready==TRUE)
      printf("FT_OUT:%+13e  ",cnvt_to_double(FT_out));
    if (nonFT_Output_Ready==TRUE)
      printf("nonFT_OUT:%+13e\n",cnvt_to_double(nonFT_out));

  } /* END OF MAIN LOOP */

  if (sav_state[0]=='Y' || sav_state[0]=='y') {
    printf("Saving processors to state files FORM1_P#.SIM ...\n");
    err=fsp_svall("form1_p");
    if (err) printf("%ERROR% Error saving processor state files.\n");
  }

}        /* end of main pgm */
```

```
PE_to_PE(pe_src,pe_dst)        /* interprocessor comm function */
                               /* SEE PE->PE interface design doc. for background */
int pe_src,pe_dst;
{
 unsigned long pdata,pforce;
 int tmp,s1,s0;

  fsp_wpin(pe_dst,PIN_aha,PINVAL_H);                /* hold HA* port A high on dst */

  tmp=fsp_rpin(pe_src,PIN_bts);                     /* feed TS* port B on source */
  fsp_wpin(pe_dst,PIN_ats,tmp);                     /* to TS* port A input on dst */

  tmp=fsp_rpin(pe_src,PIN_brw);                     /* feed R/W* port B on source */
  fsp_wpin(pe_dst,PIN_arw,tmp);                     /* to R/W* port A input on dst */

   fsp_rport(pe_src,PORT_BA,&pdata,&pforce);        /* read port B addr bus */
                                                    /* on src */
  if (pforce!=IDLE)                                 /* if any valid data present */
    fsp_wport(pe_dst,PORT_AA, pdata, pforce);       /* write to port A addr bus */
                                                    /* on dst */

  s1=fsp_rpin(pe_src,PIN_bs1);                      /* read src port B S1 pin */
  s0=fsp_rpin(pe_src,PIN_bs0);                      /* read src port B S0 pin */
                                                    /* where: S1=S0=0 -> Y mem */

                                                    /* address decoder: */
  if (s1==PINVAL_L && s0==PINVAL_L && (pdata==R_ICS || pdata==R_TX))
    fsp_wpin(pe_dst,PIN_ahs,PINVAL_L);              /* assert HS* on port A of */
  else fsp_wpin(pe_dst,PIN_ahs,PINVAL_H);           /* dst iff addr=Y:R_ICS */
                                                    /* or addr=Y:R_TX */

   fsp_rport(pe_dst,PORT_AD,&pdata,&pforce);        /* read port A data bus */
                                                    /* on dst (status feedbk) */
  if (pforce!=IDLE) {                               /* if any valid data present */
    fsp_wport(pe_src,PORT_BD,pdata,pforce);         /* write to port B data bus */
  }                                                 /* on src (status feedbk) */

   fsp_rport(pe_src,PORT_BD,&pdata,&pforce);        /* read port B data bus */
                                                    /* on src */
  if (pforce!=IDLE)                                 /* if any valid data present */
    fsp_wport(pe_dst,PORT_AD,pdata,pforce);         /* write to port A data bus */
                                                    /* on dst */

   fsp_rreg(pe_dst,HST_AICS,&pdata);                /* get ICS from dst port A */
  fsp_wmem(pe_src,MEM_Y,R_ICS,&pdata);              /* return value to R_ICS loc */
                                                    /* (to override simulator's */
                                                    /* desire to treat as mem) */

}       /* end of PE_to_PE function */
```

```
double cnvt_to_double(in_val) /* Convert IEEE fp longword into double    */
unsigned long in_val;
{
/* This function converts a single-precision IEEE-754 floating point      */
/* value represented in a single unsigned longword to the floating        */
/* point format of the C compiler in double precision.  This will,        */
/* among other things, allow these IEEE values to be displayed in a       */
/* more understandable form.                                              */

/* EDIT HISTORY:                                                          */
/* ---------------------------------------------------------------------- */
/* 01/20/91 adg created                                                   */
/* 01/21/91 adg bug fixes                                                 */

/* IEEE-754 1985 standard single-precision floating-point real format     */
/*                                                                        */
/*  31 30       23 22                                      0              */
/*  +--+-----------+---------------------------------------+              */
/*  | S | 8-bit exp | 23-bit mantissa f forming 1.f result |             */
/*  +--+-----------+---------------------------------------+              */

  double mantissa,result;
  unsigned sign;
  long int exponent;

  if (in_val==0) result=0.0;                        /* handle zero separately */
  else {
    mantissa = ((double) (in_val & 0x007FFFFF));    /* mask all but mantissa */
    mantissa = 1.0 + (mantissa/0x00800000);         /* scale as appropriate */
    exponent = (in_val & 0x7F800000) >> 23;         /* mask all but exponent */
    exponent = exponent - 127;                      /* remove the bias */
    if (exponent>=0)
      result = ((double) (1 << exponent));          /* and form the value */
    else
      result = 1.0/((double) (1 << -exponent));
    result = result * mantissa;                     /* compute overall value */

    sign = (in_val & 0x80000000) >> 31;             /* mask all but sign */
    if (sign!=0) result = result * -1.0;            /* include sign if present */
  }
  return(result);
}
```

```
unsigned long cnvt_to_IEEE(in_val) /* Convert double into IEEE fp longword */
double in_val;
{
/* This function converts a floating point double value from the C        */
/* to a single-precision IEEE-754 floating point value                    */
/* represented in a single unsigned longword. This will allow, for        */
/* example, C double values to be inputed to simulations.                 */

/* EDIT HISTORY:                                                          */
/* ---------------------------------------------------------------------- */
/* 01/20/91 adg created                                                   */
/* 01/21/91 adg bug fixes                                                 */

/* IEEE-754 1985 standard single-precision floating-point real format     */
/*                                                                        */
/*  31 30      23 22                                   0                  */
/*  +--+-----------+-----------------------------------+                  */
/*  | S | 8-bit exp | 23-bit mantissa f forming 1.f result |              */
/*  +--+-----------+-----------------------------------+                  */

 unsigned sign;
 long int exponent;
 unsigned long result;

 if (in_val==0) result=0;                    /* if zero, handle separately */
 else {
   if (in_val<0) {                           /* if it is a negative value, */
    result = 0x80000000;                     /* set the sign bit */
     in_val = in_val * -1;                   /* and make value positive */
   }
   else result = 0x00000000;                 /* o/w reset the sign bit */

   exponent = 0;                             /* start with zero */
   while (in_val<1.0 || in_val>=2.0) {       /* loop until mantissa in range */
     if (in_val<1.0) {                       /* if too small */
      in_val = in_val * 2.0;                 /* double and decrement exponent */
      exponent = exponent - 1;
     }
     else {
      in_val = in_val / 2.0;                 /* or if too large */
      exponent = exponent + 1;               /* halve and decrement exponent */
     }
   }
                                             /* add exp bias and shift in place */
   result = result | ((exponent+127)<<23);

   in_val = (in_val - 1.0) * 0x00800000;     /* scale mantissa appropriately */
   result = result | ((unsigned long) in_val); /* and move mantissa in place */
 }
 return(result);
}
```

Appendix C
SYSTEM MACRO AND EQUATE SOFTWARE LISTINGS

This appendix provides a complete listing of the DSP96002 assembly-language operating system macros and equates developed for this text. This software provides the communication, digital filtering, and FFT primitives with which tests and evaluations have been undertaken.

```
        TITLE   "(OP_SYS.INC) Operating System Equates & Macros include file"
        OPT CC,MEX
        PAGE 166,47

; EDIT HISTORY:
;--------------
; 01/26/91 adg  created initial equates and macros
; 01/27/91 adg  addition of DMA equates
; 01/28/91 adg  added SYS_IN & SYS_OUT equates
;               updated 'get_PE' & 'get_IN' macros
; 01/31/91 adg  added equate (OUT_ACK) and changed macro 'put_OUT' to provide for
;               acknowledgement back to final PE in each array so as to
;               introduce an overall output-based synchronization
; 02/17/91 adg  minor modification to macro naming scheme (reflects 2 modes)
; 02/19/91 adg  added macros (init and calc) for digital filtering support
; 02/23/91 adg  repaired minor error in digital filter macros
; 02/25/91 adg  modifications supporting change from asynch to synch clocks
; 03/10/91 adg  addition of new macros & extra equates for DMA input/output
; 03/14/91 adg  continuing repairs for dma communication
; 03/16/91 adg  additions for support of internal sin/cos ROM info
; 03/16/91 adg  enhanced input dma routines for both X and Y internal dsts
; 03/23/91 adg  Removed double-buffering from DMA macros so that it is instead
;               managed by the calling application explicitly...this is needed
;               since the input data received for FFT ops goes to both X and
;               Y space (first gp of X, then gp of Y), and DMA channels can
;               only handle one at a time.
; 04/01/91 adg  Addition of multipass FFT macro (formerly in separate file).
;
;--------------------------------------------------------------------------------
; EQUATES (Internal I/O peripherals in X space; External in Y space)
;--------------------------------------------------------------------------------

                                    ; X Space Memory-Mapped Registers
IPR     equ     $FFFFFFFF           ; Interrupt Priority Register
BCRA    equ     $FFFFFFFE           ; Port A Bus Control Register
BCRB    equ     $FFFFFFFD           ; Port B Bus Control Register
PSR     equ     $FFFFFFFC           ; Port Select Register (0=A,1=B per bit)

                                    ; Host Interface X Memory-mapped regs (Port A)
HCRA    equ     $FFFFFFEC           ; Host Control Register
HSRA    equ     $FFFFFFED           ; Host Status Register (read-only)
HTXCA   equ     $FFFFFFEE           ; Host Transmit data and Clear HMRC Register
HRXA    equ     $FFFFFFEF           ; Host Receive Data Register
HTXA    equ     $FFFFFFEF           ; Host Transmit Data Register

                                    ; Host Interface X Memory-mapped regs (Port B)
HCRB    equ     $FFFFFFE4           ; Host Control Register
HSRB    equ     $FFFFFFE5           ; Host Status Register (read-only)
HTXCB   equ     $FFFFFFE6           ; Host Transmit Data and Clear HMRC Register
```

```
HRXB     equ  $FFFFFFE7   ; Host Receive Data Register
HTXB     equ  $FFFFFFE7   ; Host Transmit Data Register

                          ; X space memory-mapped DMA Channel 0 registers
DSM0     equ  $FFFFFFDF   ; DMA Ch0 Source Modifier Register
DSR0     equ  $FFFFFFDE   ; DMA Ch0 Source Address Register
DSN0     equ  $FFFFFFDD   ; DMA Ch0 Source Offset Register
DCO0     equ  $FFFFFFDC   ; DMA Ch0 Counter Register
DDM0     equ  $FFFFFFDB   ; DMA Ch0 Destination Modifier Register
DDR0     equ  $FFFFFFDA   ; DMA Ch0 Destination Address Register
DDN0     equ  $FFFFFFD9   ; DMA Ch0 Destination Offset Register
DCS0     equ  $FFFFFFD8   ; DMA Ch0 Control/Status Register

                          ; X space memory-mapped DMA Channel 1 registers
DSM1     equ  $FFFFFFD7   ; DMA Ch1 Source Modifier Register
DSR1     equ  $FFFFFFD6   ; DMA Ch1 Source Address Register
DSN1     equ  $FFFFFFD5   ; DMA Ch1 Source Offset Register
DCO1     equ  $FFFFFFD4   ; DMA Ch1 Counter Register
DDM1     equ  $FFFFFFD3   ; DMA Ch1 Destination Modifier Register
DDR1     equ  $FFFFFFD2   ; DMA Ch1 Destination Address Register
DDN1     equ  $FFFFFFD1   ; DMA Ch1 Destination Offset Register
DCS1     equ  $FFFFFFD0   ; DMA Ch1 Control/Status Register

                          ; Host Functions (mapped at end of Y space)
                          ; Used in communicating w/ remote DSP96Ks.
                          ; These are just one of many possible sets
                          ; that could be used; the key points are that
                          ; bits A5-A2 represent the host func. code &
                          ; the addr overall is in Y ext periph I/O space
R_ICS    equ  $FFFFFFE0   ; ICS register r/w (host func. 1000B)
R_SEM    equ  $FFFFFFE4   ; SEM register r/w (host func. 1001B)
R_RX     equ  $FFFFFFE8   ; RX register read (host func. 1010B)
R_TX     equ  $FFFFFFE8   ; TX register write (host func. 1010B)
R_IVR    equ  $FFFFFFF0   ; IVR register r/w (host func. 1100B)
R_CVR    equ  $FFFFFFF4   ; CVR register r/w (host func. 1101B)
R_TXYW   equ  $FFFFFFC0   ; TX reg write and Y mem write (func. 0000B)
R_TXYR   equ  $FFFFFFC4   ; TX reg write and Y mem read (func. 0001B)
R_TXXW   equ  $FFFFFFC8   ; TX reg write and X mem write (func. 0010B)
R_TXXR   equ  $FFFFFFCC   ; TX reg write and X mem read (func. 0011B)
R_TXPW   equ  $FFFFFFD0   ; TX reg write and P mem write (func. 0100B)
R_TXPR   equ  $FFFFFFD4   ; TX reg write and P mem read (func. 0101B)

                          ; Individual Bit Positions
HTDE     equ  1           ; Host Transmit Data Empty bit in each HSR
HRDF     equ  0           ; Host Receive Data Full bit in each HSR
TXDE     equ  1           ; TX Data Empty bit in each ICS
RXDF     equ  0           ; RX Data Full bit in each ICS
INIT     equ  6           ; INIT HI Initialize bit in each ICS
DTD      equ  28          ; DMA Transfer Done bit in each DCS
```

```
                                        ; Special Memory-Mapped Addresses for In->PE
                                        ; and PE->Output communication
SYS_IN      equ     $FFFFFF7F           ; X space external memory addr (for input)
SYS_OUT     equ     $FFFFFF80           ; Y space external memory addr (for output)

ROM_TBL     equ     $00000400           ; Start addr of X/Y cos/sin ROM tbl data
ROM_LEN     equ     $00000400           ; Length of each ROM cos/sin table (# pts)

;-------------------------------------------------------------------------------
; MACROS
;-------------------------------------------------------------------------------

init_PE     macro                               ; perform all necessary register
                                                ; initializations for this PE

            move    #$0000FF00,X:PSR            ; map external Y refs to port B,
                                                ; X refs to A, and P refs also to A

            move    #8,OMR                      ; set DE bit in OMR to enable ROMs
            endm

;-------------------------------------------------------------------------------

put_PE      macro   REG                         ; macro to perform put operation over
                                                ; host i/f port B to the next PE in
                                                ; the array when it is ready
                                                ; e.g. called via 'put_PE D0.S'

_LP1        JCLR    #TXDE,Y:R_ICS,_LP1          ; loop while TXDE bit of the remote
                                                ; ICS reg on the slave is zero, via
                                                ; host function 1000 for ICS read

            MOVEP   REG,Y:R_TX                  ; now that the remote PE is ready to
                                                ; receive, go ahead and send via
                                                ; host function TX register write
                                                ; by writing to a local Y ext peripheral
                                                ; loc memory-mapped to TX on the slave
            endm

;-------------------------------------------------------------------------------

put_OUT     macro   REG                         ; macro to perform put operation over
                                                ; host i/f port B to output latch
                                                ; e.g. called via 'put_OUT D0.S'

            MOVE    REG,Y:SYS_OUT               ; move register to output via
                                                ; memory write (treating the output as
```

```
                                                ; a memory element) @ a special
address
        endm

;-----------------------------------------------------------------------------------

get_PE  macro   REG                             ; macro to perform get operation over
                                                ; host i/f port A from the previous PE
                                                ; in the array when it is ready
                                                ; e.g. called via 'get_PE D0.S'

_LP1    JCLR    #HRDF,X:HSRA,_LP1               ; loop while HRDF bit of the local
                                                ; HSR reg is zero on port A

        MOVEP   X:HRXA,REG                      ; now that the data has been received
                                                ; by the host interface, go ahead
                                                ; and read it via the local HRX reg
                                                ; for port A
        endm

;-----------------------------------------------------------------------------------

get_IN  macro   REG                             ; macro to perform get operation from
                                                ; the system input (FT or o/w)
                                                ; located previous in the array
                                                ; e.g. called via 'get_IN D0.S'

        MOVE    X:SYS_IN,REG                    ; move input to register via
                                                ; memory read (treating the input as
                                                ; a memory element) @ a special
address
        endm

;-----------------------------------------------------------------------------------
;-----------------------------------------------------------------------------------

init_DF macro   B,DATA,COEFF            ; digital filter initialization macro.
                                        ; B indicates the # of biquads to be done,
                                        ; while DATA and COEFF point to the delay
                                        ; data array in X space and the coeff array
                                        ; in Y space respectively.

; Implements B cascaded real biquad IIR filters
; Digital Filter is of form 1D:
; w(n) = x(n) - a1*w(n-1) - a2*w(n-2), y(n) = w(n) + b1*w(n-1) + b2*w(n-2)
; where: x(n) is input, y(n) is output, and w(n) is temporary

        move    #(2*B-1),m0     ; init delay data mod regs for modulo-2B access
        move    m0,m1
```

```
        move     #(4*B-1),m4        ; init coeff mod reg for modulo-4B access
        move     #DATA,r0           ; init addr reg to read from delayed data array
        move     #COEFF,r4          ; init addr reg to coeff array
        move     r0,r1              ; init addr reg to write to delayed data array

        ; clear tmp reg, get initial w(n-2) data, & get initial a2 coeff
        fclr     d1                 x:(r0)+,d4.s  y:(r4)+,d6.s

        endm

;-----------------------------------------------------------------------------------------

calc_DF macro    B,REG                  ; calculate the next digital filter output.
                                        ; B indicates the # of biquads calculated,
                                        ; while REG is a register which provides
                                        ; the input and receives the output

; e.g. for B=2, calculations carry out the following equations:
; w1(n) = x(n)  - a11*w1(n-1) - a21*w1(n-2)
; y1(n) = w1(n) - b11*w1(n-1) + b21*w1(n-2)
; w2(n) = y1(n) - a12*w2(n-1) - a22*w2(n-1)
; y2(n) = w2(n) + b12*w2(n-1) + b22*w2(n-2) = y(n)

        do       #B,ENDDO           ; loop N times

        ; start calc of w(n) and finish that of previous y(n):
        ; do a2*w(n-2), add b1*w(n-1) from previous, get w(n-1), & get a1 coeff
        fmpy     d4,d6,d1  fadd.s d1,REG        x:(r0)+,d5.s  y:(r4)+,d6.s

        ; cont calc of w(n) and delay data for next loop:
        ; do a1*w(n-1), sub a2*w(n-2) from subtot, w(n-1)->w(n-2), & get b2
        fmpy     d5,d6,d1  fsub.s d1,REG        d5.s,x:(r1)+  y:(r4)+,d6.s

        ; start calc of y(n), finish w(n), & prepare for next w(n):
        ; do b2*w(n-2), sub a1*w(n-1) from subtot, get next w(n-1), & get b1
        fmpy     d4,d6,d1  fsub.s d1,REG        x:(r0)+,d4.s  y:(r4)+,d6.s

        ; cont calc of y(n), delay data for next loop, & prep for next w(n):
        ; do b1*w(n-1), add b2*w(n-2) from subtot, w(n)->w(n-1), & get next a2
        fmpy     d5,d6,d1  fadd.s d1,REG        REG,x:(r1)+  y:(r4)+,d6.s
ENDDO
        fadd.s   d1,REG                 ; add final b1*w(n-1) to final result y(n)
        fclr     d1                     ; clear tmp register in case another invocation

        endm

;-----------------------------------------------------------------------------------------
;-----------------------------------------------------------------------------------------
```

```
mpfft     macro     POINTS,DATA,COEFF,COEFFSIZE,START_PASS,NUM_PASSES
;
; This macro is based on code in [DSP89] for a Radix-2 DIT In-Place FFT, where:
;   - uses complex input and output data (real in X, imag in Y)
;   - normally ordered input data
;   - bit reversed output data
;   - POINTS = # of points (any pwr of 2 in range 2<->2G) = N
;   - DATA = start of data buffer
;   - COEF = start of 1/2 cycle cosine/sine table
;   - COEFSIZE = number of table pts in cosine/sine table (one of
;                     multiples pts/2, pts, 3*pts/2,2*pts,5*pts/2, etc.)
;   - START_PASS = no. of the first FFT pass to be executed
;   - NUM_PASSES = no. of FFT passes to be executed
; (the last two parameters are used to provide selective execution of a
;  subset of all passes...this makes it possible for implementation on
;  the FT real-time DSP multiprocessor being developed via stages)
;
; There are Log2(N) passes for this FFT, and during each pass the data is
; fetched from memory, used to perform the bfly calculations, and the result
; is stored back to memory. Within each pass, bly's are clustered into groups,
; where the #gps/pass doubles for each pass (starting at 1) and the #bly's/gp
; halves for each pass (starting at N/2). The twiddle factors are the same
; for all bly's in each group, and the order of twiddle factors from one gp
; to the next is bit-reversed.

          move    #POINTS,d1.l                             ; store N for FFT
          move    #@cvi(@log(POINTS)/@log(2)+0.5),n1      ; store Log2(N) pass cnt
          move    #DATA,r2                                 ; store start addr of data
          move    #COEFF,m2                      ; store " addr of coef tbl
          move    #COEFFSIZE,d2.l                ; store # pts in 1/2 cycle

          move    #0,m6                          ; reverse carry (bit-reversed update)
          move    #-1,m0                         ; set m0 for linear update (via r0)
          clr     d0              m0,m1          ; set m1 for linear update, set d0 to 1
          inc     d0              m0,m4          ; for init gp cnt, & set m4 for lin upd
          lsr     d2              m0,m5          ; halve d2 and set m5 for linear upd
          move    d2.l,n6                        ; use 1/2 coeff size as bit-rev offset

          IF      START_PASS>1                   ; if starting beyond the 1st pass
          do      #(START_PASS-1),_prep          ; loop for missed initial pass(es)
          lsr     d1                             ; update (halve) bfly cnt
          lsl     d0                             ; update (double) group cnt
_prep
          ENDIF                                  ; end of conditional assembly

          do      #NUM_PASSES,_passend           ; loop for all requested passes
          move    r2,r0                          ; set r0 to also point to data
          move    d0.l,n2                        ; store gp cnt into reg for loop cntr
          lsr     d1              m2,r6          ; calc bfly cnt & point r6 to coeff tbl
```

```
        dec      d1                  d1.l,n0    ; store bfly cnt (initially N/2)
        move     d1.l,n1
        move     n0,n4                          ; use same offset inc for n4 as n0
        move     n0,n5                          ; as well as for n5
        lea      (r0)+n0,r1
        lea      (r0)-,r4
        lea      (r1)-,r5

        do       n2,_grpend                     ; loop for each gp in this pass
                                                ; (# gps/pass = 1 for 1st pass, &
                                                ;  doubles for each pass afterwards)
                                                ; do prep for the bfly calculations:
        move                                    x:(r6)+n6,d9.s  y:,d8.s
        move                                    x:(r1)+,d6.s    y:,d7.s
        fmpy.s   d8,d7,d3                                       y:(r5),d2.s
        fmpy.s   d9,d6,d0                                       y:(r4),d5.s
        fmpy.s   d9,d7,d1                                       y:(r1),d7.s

        ; Summary of Calculations Performed in FFT Butterfly Code:
        ; ar' = ar + (wr*br + wi*bi) = ar + x
        ; ai' = ai + (wr*bi - wi*br) = ai - y
        ; br' = ar - (wr*br + wi*bi) = ar - x
        ; bi' = ai - (wr*bi + wi*br) = ai + y

        do       n0,_bflyend                ; loop for all bfly's in this group
                                            ; (# bly's/gp = N/2 for 1st pass, &
                                            ;  halves for each pass afterwards)
        ;        d2=d8*d6   d0=d0+d3
        ;         =wi*br     =x=wr*br+wi*bi      load d4=ar      store d2=bi'
        fmpy     d8,d6,d2   fadd.s   d3,d0       x:(r0),d4.s     d2.s,y:(r5)+

        ;        d3=d8*d7   d4=d4-d0=br'=ar-x
        ;         =wi*bi    d0=d4+d0=ar'=ar+x    load d6=br      store d5=ai'
        fmpy     d8,d7,d3   faddsub.s d4,d0      x:(r1)+,d6.s    d5.s,y:(r4)+

        ;        d0=d9*d8   d2=d2-d1
        ;         =wr*br     =y=wi*br-wr*bi      store d0=ar'    load d5=ai
        fmpy     d9,d6,d0   fsub.s   d1,d2       d0.s,x:(r4)     y:(r0)+,d5.s

        ;        d1=d9*d7   d5=d5-d2=ai'=ai-y
        ;         =wr*bi    d2=d5+d2=bi'=ai+y    store d4=br'    load d7=bi
        fmpy     d9,d7,d1   faddsub.s d5,d2      d4.s,x:(r5)     y:(r1),d7.s

_bflyend
                                                ; prepare for next group:
        move                                    x:(r0)+n0,d7.s  d2.s,y:(r5)+n5
        move                                    x:(r1)+n1,d7.s  d5.s,y:(r4)+n4
_grpend                                         ; prepare for next pass:
        move     n2,d0.l                        ; restore gp cnt
```

```
        lsl     d0          n0,d1.l     ; dble gp cnt for next pass
                                        ; and restore bfly cnt
_passend                                ; all passes complete!
        endm
```

Appendix D
PRELIMINARY TEST SOFTWARE LISTINGS

This appendix provides DSP96002 assembly-language listings for the actual preliminary test programs developed for this text. Those programs include single-word data propagation tests, multiple-word data propagation tests, digital filtering tests, and FFT tests.

```
        title     "(PROP1.ASM) data propagation test for initial stage PEs"

        include   'OP_SYS.INC'   ; include operating system equates & macros

        org       p:$100
START
        init_PE                  ; initialize this PE

LP      get_IN    D0.S           ; get system input
        put_PE    D0.S           ; transfer to next PE in array

        JMP       LP             ; continue infinite loop
        STOP
```

```
        title     "(PROP2.ASM) data propagation test for intermediate stage PEs"

        include   'OP_SYS.INC'    ; include operating system equates & macros

        org       p:$100
START
        init_PE                   ; initialize this PE

LP      get_PE    D0.S            ; get input from previous PE in array
        put_PE    D0.S            ; transfer to next PE in array

        JMP       LP              ; continue infinite loop
        STOP
```

```
        title     "(PROP3.ASM) data propagation test for final stage PEs"

        include   'OP_SYS.INC'    ; include operating system equates & macros

        org       p:$100
START
        init_PE                   ; initialize this PE

LP      get_PE    D0.S            ; get input from previous PE in array
        put_OUT   D0.S            ; transfer to the output latch

        JMP       LP              ; continue infinite loop
        STOP
```

```
        title     "(MP8_1.ASM) multiword data prop test for initial stage PEs"

        include   'OP_SYS.INC'     ; incl operating system equates & macros

N       equ       8
        org       x:$0
buff    ds        N                ; set aside buffer space for N pts (real)
        org       y:$0
        ds        N                ; set aside buffer space for N pts (imag)

        org       p:$100
START
        init_PE                    ; init this PE
        move      #-1,m0           ; linear access
        move      #1,n0            ; w/ offset inc of 1
;       move      #0,m0            ; bit-reverse access
;       move      #(N/2),n0        ; w/ offset inc associated w/ bit-rev access

LP      move      #buff,r0         ; initialize input data ptr register
        do        #N,lp1           ; transfer each of the input values
        get_IN    d0.s             ; from previous input/PE.  First loop
        move      d0.s,x:(r0)+n0   ; for all of the real parts of the results,
lp1
        move      #buff,r0         ; initialize input data ptr register
        do        #N,lp2           ; then loop for all of the imag parts
        get_IN    d0.s             ; (both with normal output ordering)
        move      d0.s,y:(r0)+n0
lp2
                                   ; <INSERT DATA PROCESSING CODE HERE>

        move      #buff,r0         ; initialize output data ptr register
        do        #N,lp3           ; transfer each of the output results
        nop                        ; to next PE in the array.  First loop
        put_PE    x:(r0)+n0        ; for all of the real parts of the results,
        nop                        ; (NOPs for DO instruction restrictions)
lp3
        move      #buff,r0         ; initialize output data ptr register
        do        #N,lp4           ; then loop for all of the imag parts
        nop                        ; (both with normal output ordering)
        put_PE    y:(r0)+n0
        nop                        ; (NOPs for DO instruction restrictions)
lp4
        jmp       LP               ; continue infinite loop
        stop
        end       START
```

```
        title     "(MP8_2.ASM) multiword data prop test for intermed stage PEs"

        include   'OP_SYS.INC'      ; incl operating system equates & macros

N       equ       8
        org       x:$0
buff    ds        N                 ; set aside buffer space for N pts (real)
        org       y:$0
        ds        N                 ; set aside buffer space for N pts (imag)

        org       p:$100
START
        init_PE                     ; init this PE
        move      #-1,m0            ; linear access
        move      #1,n0             ; w/ offset inc of 1
;       move      #0,m0             ; bit-reverse access
;       move      #(N/2),n0         ; w/ offset inc associated w/ bit-rev access

LP      move      #buff,r0          ; initialize input data ptr register
        do        #N,lp1            ; transfer each of the input values
        nop                         ; from previous input/PE. First loop
        get_PE    d0.s              ; for all of the real parts of the results,
        move      d0.s,x:(r0)+n0
        nop                         ; (NOPs for DO instruction restrictions)
lp1
        move      #buff,r0          ; initialize input data ptr register
        do        #N,lp2            ; then loop for all of the imag parts
        nop                         ; (both with normal output ordering)
        get_PE    d0.s
        move      d0.s,y:(r0)+n0
        nop                         ; (NOPs for DO instruction restrictions)
lp2
                                    ; <INSERT DATA PROCESSING CODE HERE>

        move      #buff,r0          ; initialize output data ptr register
        do        #N,lp3            ; transfer each of the output results
        nop                         ; to next PE in the array. First loop
        put_PE    x:(r0)+n0         ; for all of the real parts of the results,
        nop                         ; (NOPs for DO instruction restrictions)
lp3
        move      #buff,r0          ; initialize output data ptr register
        do        #N,lp4            ; then loop for all of the imag parts
        nop                         ; (both with normal output ordering)
        put_PE    y:(r0)+n0
        nop                         ; (NOPs for DO instruction restrictions)
lp4
        jmp       LP                ; continue infinite loop
        stop
        end       START
```

```
          title      "(MP8_3.ASM) multiword data prop test for final stage PEs"

          include    'OP_SYS.INC'     ; incl operating system equates & macros

N         equ        8
          org        x:$0
buff      ds         N                ; set aside buffer space for N pts (real)
          org        y:$0
          ds         N                ; set aside buffer space for N pts (imag)

          org        p:$100
START
          init_PE                     ; init this PE
          move       #-1,m0           ; linear access
          move       #1,n0            ; w/ offset inc of 1
;         move       #0,m0            ; bit-reverse access
;         move       #(N/2),n0        ; w/ offset inc associated w/ bit-rev access

LP        move       #buff,r0         ; initialize input data ptr register
          do         #N,lp1           ; transfer each of the input values
          nop                         ; from previous input/PE.  First loop
          get_PE     d0.s             ; for all of the real parts of the results,
          move       d0.s,x:(r0)+n0
          nop                         ; (NOPs for DO instruction restrictions)
lp1
          move       #buff,r0         ; initialize input data ptr register
          do         #N,lp2           ; then loop for all of the imag parts
          nop                         ; (both with normal output ordering)
          get_PE     d0.s
          move       d0.s,y:(r0)+n0
          nop                         ; (NOPs for DO instruction restrictions)
lp2
                                      ; <INSERT DATA PROCESSING CODE HERE>

          move       #buff,r0         ; initialize output data ptr register
          do         #N,lp3           ; transfer each of the output results
          put_OUT    x:(r0)+n0        ; to next PE in the array.  First loop
lp3                                   ; for all of the real parts of the results,

          move       #buff,r0         ; initialize output data ptr register
          do         #N,lp4           ; then loop for all of the imag parts
          put_OUT    y:(r0)+n0        ; (both with normal output ordering)
lp4
          jmp        LP               ; continue infinite loop
          stop
          end        START
```

```
        title       "(MP16_1.ASM) multiword data prop test for initial stage PEs"

        include     'OP_SYS.INC'       ; incl operating system equates & macros

N       equ         16
        org         x:$0
buff    ds          N                  ; set aside buffer space for N pts (real)
        org         y:$0
        ds          N                  ; set aside buffer space for N pts (imag)

        org         p:$100
START
        init_PE                        ; init this PE
        move        #-1,m0             ; linear access
        move        #1,n0              ; w/ offset inc of 1
;       move        #0,m0              ; bit-reverse access
;       move        #(N/2),n0          ; w/ offset inc associated w/ bit-rev access

LP      move        #buff,r0           ; initialize input data ptr register
        do          #N,lp1             ; transfer each of the input values
        get_IN      d0.s               ; from previous input/PE.  First loop
        move        d0.s,x:(r0)+n0     ; for all of the real parts of the results,
lp1
        move        #buff,r0           ; initialize input data ptr register
        do          #N,lp2             ; then loop for all of the imag parts
        get_IN      d0.s               ; (both with normal output ordering)
        move        d0.s,y:(r0)+n0
lp2
                                       ; <INSERT DATA PROCESSING CODE HERE>

        move        #buff,r0           ; initialize output data ptr register
        do          #N,lp3             ; transfer each of the output results
        nop                            ; to next PE in the array.  First loop
        put_PE      x:(r0)+n0          ; for all of the real parts of the results,
        nop                            ; (NOPs for DO instruction restrictions)
lp3
        move        #buff,r0           ; initialize output data ptr register
        do          #N,lp4             ; then loop for all of the imag parts
        nop                            ; (both with normal output ordering)
        put_PE      y:(r0)+n0
        nop                            ; (NOPs for DO instruction restrictions)
lp4
        jmp         LP                 ; continue infinite loop
        stop
        end         START
```

```
        title     "(MP16_2.ASM) multiword data prop test for intermed stage PEs"

        include   'OP_SYS.INC'    ; incl operating system equates & macros

N       equ       16
        org       x:$0
buff    ds        N               ; set aside buffer space for N pts (real)
        org       y:$0
        ds        N               ; set aside buffer space for N pts (imag)

        org       p:$100
START
        init_PE                   ; init this PE
        move      #-1,m0          ; linear access
        move      #1,n0           ; w/ offset inc of 1
;       move      #0,m0           ; bit-reverse access
;       move      #(N/2),n0       ; w/ offset inc associated w/ bit-rev access

LP      move      #buff,r0        ; initialize input data ptr register
        do        #N,lp1          ; transfer each of the input values
        nop                       ; from previous input/PE. First loop
        get_PE    d0.s            ; for all of the real parts of the results,
        move      d0.s,x:(r0)+n0
        nop                       ; (NOPs for DO instruction restrictions)
lp1
        move      #buff,r0        ; initialize input data ptr register
        do        #N,lp2          ; then loop for all of the imag parts
        nop                       ; (both with normal output ordering)
        get_PE    d0.s
        move      d0.s,y:(r0)+n0
        nop                       ; (NOPs for DO instruction restrictions)
lp2
                                  ; <INSERT DATA PROCESSING CODE HERE>

        move      #buff,r0        ; initialize output data ptr register
        do        #N,lp3          ; transfer each of the output results
        nop                       ; to next PE in the array. First loop
        put_PE    x:(r0)+n0       ; for all of the real parts of the results,
        nop                       ; (NOPs for DO instruction restrictions)
lp3
        move      #buff,r0        ; initialize output data ptr register
        do        #N,lp4          ; then loop for all of the imag parts
        nop                       ; (both with normal output ordering)
        put_PE    y:(r0)+n0
        nop                       ; (NOPs for DO instruction restrictions)
lp4
        jmp       LP              ; continue infinite loop
        stop
        end       START
```

```
        title     "(MP16_3.ASM) multiword data prop test for final stage PEs"

        include   'OP_SYS.INC'    ; incl operating system equates & macros

N       equ       16
        org       x:$0
buff    ds        N               ; set aside buffer space for N pts (real)
        org       y:$0
        ds        N               ; set aside buffer space for N pts (imag)

        org       p:$100
START
        init_PE                   ; init this PE
        move      #-1,m0          ; linear access
        move      #1,n0           ; w/ offset inc of 1
;       move      #0,m0           ; bit-reverse access
;       move      #(N/2),n0       ; w/ offset inc associated w/ bit-rev access

LP      move      #buff,r0        ; initialize input data ptr register
        do        #N,lp1          ; transfer each of the input values
        nop                       ; from previous input/PE.  First loop
        get_PE    d0.s            ; for all of the real parts of the results,
        move      d0.s,x:(r0)+n0
        nop                       ; (NOPs for DO instruction restrictions)
lp1
        move      #buff,r0        ; initialize input data ptr register
        do        #N,lp2          ; then loop for all of the imag parts
        nop                       ; (both with normal output ordering)
        get_PE    d0.s
        move      d0.s,y:(r0)+n0
        nop                       ; (NOPs for DO instruction restrictions)
lp2
                                  ; <INSERT DATA PROCESSING CODE HERE>

        move      #buff,r0        ; initialize output data ptr register
        do        #N,lp3          ; transfer each of the output results
        put_OUT   x:(r0)+n0       ; to next PE in the array.  First loop
lp3                               ; for all of the real parts of the results,

        move      #buff,r0        ; initialize output data ptr register
        do        #N,lp4          ; then loop for all of the imag parts
        put_OUT   y:(r0)+n0       ; (both with normal output ordering)
lp4
        jmp       LP              ; continue infinite loop
        stop
        end       START
```

```
        title       "(MP32_1.ASM) multiword data prop test for initial stage PEs"

        include     'OP_SYS.INC'        ; incl operating system equates & macros

N       equ         32
        org         x:$0
buff    ds          N                   ; set aside buffer space for N pts (real)
        org         y:$0
        ds          N                   ; set aside buffer space for N pts (imag)

        org         p:$100
START
        init_PE                         ; init this PE
        move        #-1,m0              ; linear access
        move        #1,n0               ; w/ offset inc of 1
;       move        #0,m0               ; bit-reverse access
;       move        #(N/2),n0           ; w/ offset inc associated w/ bit-rev access

LP      move        #buff,r0            ; initialize input data ptr register
        do          #N,lp1              ; transfer each of the input values
        get_IN      d0.s                ; from previous input/PE. First loop
        move        d0.s,x:(r0)+n0      ; for all of the real parts of the results,
lp1
        move        #buff,r0            ; initialize input data ptr register
        do          #N,lp2              ; then loop for all of the imag parts
        get_IN      d0.s                ; (both with normal output ordering)
        move        d0.s,y:(r0)+n0
lp2
                                        ; <INSERT DATA PROCESSING CODE HERE>

        move        #buff,r0            ; initialize output data ptr register
        do          #N,lp3              ; transfer each of the output results
        nop                             ; to next PE in the array. First loop
        put_PE      x:(r0)+n0           ; for all of the real parts of the results,
        nop                             ; (NOPs for DO instruction restrictions)
lp3
        move        #buff,r0            ; initialize output data ptr register
        do          #N,lp4              ; then loop for all of the imag parts
        nop                             ; (both with normal output ordering)
        put_PE      y:(r0)+n0
        nop                             ; (NOPs for DO instruction restrictions)
lp4
        jmp         LP                  ; continue infinite loop
        stop
        end         START
```

```
        title    "(MP32_2.ASM) multiword data prop test for intermed stage PEs"

        include  'OP_SYS.INC'     ; incl operating system equates & macros

N       equ      32
        org      x:$0
buff    ds       N                ; set aside buffer space for N pts (real)
        org      y:$0
        ds       N                ; set aside buffer space for N pts (imag)

        org      p:$100
START
        init_PE                   ; init this PE
        move     #-1,m0           ; linear access
        move     #1,n0            ; w/ offset inc of 1
;       move     #0,m0            ; bit-reverse access
;       move     #(N/2),n0        ; w/ offset inc associated w/ bit-rev access

LP      move     #buff,r0         ; initialize input data ptr register
        do       #N,lp1           ; transfer each of the input values
        nop                       ; from previous input/PE.  First loop
        get_PE   d0.s             ; for all of the real parts of the results,
        move     d0.s,x:(r0)+n0
        nop                       ; (NOPs for DO instruction restrictions)
lp1
        move     #buff,r0         ; initialize input data ptr register
        do       #N,lp2           ; then loop for all of the imag parts
        nop                       ; (both with normal output ordering)
        get_PE   d0.s
        move     d0.s,y:(r0)+n0
        nop                       ; (NOPs for DO instruction restrictions)
lp2
                                  ; <INSERT DATA PROCESSING CODE HERE>

        move     #buff,r0         ; initialize output data ptr register
        do       #N,lp3           ; transfer each of the output results
        nop                       ; to next PE in the array.  First loop
        put_PE   x:(r0)+n0        ; for all of the real parts of the results,
        nop                       ; (NOPs for DO instruction restrictions)
lp3
        move     #buff,r0         ; initialize output data ptr register
        do       #N,lp4           ; then loop for all of the imag parts
        nop                       ; (both with normal output ordering)
        put_PE   y:(r0)+n0
        nop                       ; (NOPs for DO instruction restrictions)
lp4
        jmp      LP               ; continue infinite loop
        stop
        end      START
```

```
            title       "(MP32_3.ASM) multiword data prop test for final stage PEs"

            include     'OP_SYS.INC'    ; incl operating system equates & macros

N           equ         32
            org         x:$0
buff        ds          N               ; set aside buffer space for N pts (real)
            org         y:$0
            ds          N               ; set aside buffer space for N pts (imag)

            org         p:$100
START
            init_PE                     ; init this PE
            move        #-1,m0          ; linear access
            move        #1,n0           ; w/ offset inc of 1
;           move        #0,m0           ; bit-reverse access
;           move        #(N/2),n0       ; w/ offset inc associated w/ bit-rev access

LP          move        #buff,r0        ; initialize input data ptr register
            do          #N,lp1          ; transfer each of the input values
            nop                         ; from previous input/PE.  First loop
            get_PE      d0.s            ; for all of the real parts of the results,
            move        d0.s,x:(r0)+n0
            nop                         ; (NOPs for DO instruction restrictions)
lp1
            move        #buff,r0        ; initialize input data ptr register
            do          #N,lp2          ; then loop for all of the imag parts
            nop                         ; (both with normal output ordering)
            get_PE      d0.s
            move        d0.s,y:(r0)+n0
            nop                         ; (NOPs for DO instruction restrictions)
lp2
                                        ; <INSERT DATA PROCESSING CODE HERE>

            move        #buff,r0        ; initialize output data ptr register
            do          #N,lp3          ; transfer each of the output results
            put_OUT     x:(r0)+n0       ; to next PE in the array.  First loop
lp3                                     ; for all of the real parts of the results,

            move        #buff,r0        ; initialize output data ptr register
            do          #N,lp4          ; then loop for all of the imag parts
            put_OUT     y:(r0)+n0       ; (both with normal output ordering)
lp4
            jmp         LP              ; continue infinite loop
            stop
            end         START
```

```
        title     "(MP64_1.ASM) multiword data prop test for initial stage PEs"

        include   'OP_SYS.INC'      ; incl operating system equates & macros

N       equ       64
        org       x:$0
buff    ds        N                 ; set aside buffer space for N pts (real)
        org       y:$0
        ds        N                 ; set aside buffer space for N pts (imag)

        org       p:$100
START
        init_PE                     ; init this PE
        move      #-1,m0            ; linear access
        move      #1,n0             ; w/ offset inc of 1
;       move      #0,m0             ; bit-reverse access
;       move      #(N/2),n0         ; w/ offset inc associated w/ bit-rev access

LP      move      #buff,r0          ; initialize input data ptr register
        do        #N,lp1            ; transfer each of the input values
        get_IN    d0.s              ; from previous input/PE.  First loop
        move      d0.s,x:(r0)+n0    ; for all of the real parts of the results,
lp1
        move      #buff,r0          ; initialize input data ptr register
        do        #N,lp2            ; then loop for all of the imag parts
        get_IN    d0.s              ; (both with normal output ordering)
        move      d0.s,y:(r0)+n0
lp2
                                    ; <INSERT DATA PROCESSING CODE HERE>

        move      #buff,r0          ; initialize output data ptr register
        do        #N,lp3            ; transfer each of the output results
        nop                         ; to next PE in the array.  First loop
        put_PE    x:(r0)+n0         ; for all of the real parts of the results,
        nop                         ; (NOPs for DO instruction restrictions)
lp3
        move      #buff,r0          ; initialize output data ptr register
        do        #N,lp4            ; then loop for all of the imag parts
        nop                         ; (both with normal output ordering)
        put_PE    y:(r0)+n0
        nop                         ; (NOPs for DO instruction restrictions)
lp4
        jmp       LP                ; continue infinite loop
        stop
        end       START
```

```
        title     "(MP64_2.ASM) multiword data prop test for intermed stage PEs"

        include   'OP_SYS.INC'    ; incl operating system equates & macros

N       equ       64
        org       x:$0
buff    ds        N               ; set aside buffer space for N pts (real)
        org       y:$0
        ds        N               ; set aside buffer space for N pts (imag)

        org       p:$100
START
        init_PE                   ; init this PE
        move      #-1,m0          ; linear access
        move      #1,n0           ; w/ offset inc of 1
;       move      #0,m0           ; bit-reverse access
;       move      #(N/2),n0       ; w/ offset inc associated w/ bit-rev access

LP      move      #buff,r0        ; initialize input data ptr register
        do        #N,lp1          ; transfer each of the input values
        nop                       ; from previous input/PE.  First loop
        get_PE    d0.s            ; for all of the real parts of the results,
        move      d0.s,x:(r0)+n0
        nop                       ; (NOPs for DO instruction restrictions)
lp1
        move      #buff,r0        ; initialize input data ptr register
        do        #N,lp2          ; then loop for all of the imag parts
        nop                       ; (both with normal output ordering)
        get_PE    d0.s
        move      d0.s,y:(r0)+n0
        nop                       ; (NOPs for DO instruction restrictions)
lp2
                                  ; <INSERT DATA PROCESSING CODE HERE>

        move      #buff,r0        ; initialize output data ptr register
        do        #N,lp3          ; transfer each of the output results
        nop                       ; to next PE in the array.  First loop
        put_PE    x:(r0)+n0       ; for all of the real parts of the results,
        nop                       ; (NOPs for DO instruction restrictions)
lp3
        move      #buff,r0        ; initialize output data ptr register
        do        #N,lp4          ; then loop for all of the imag parts
        nop                       ; (both with normal output ordering)
        put_PE    y:(r0)+n0
        nop                       ; (NOPs for DO instruction restrictions)
lp4
        jmp       LP              ; continue infinite loop
        stop
        end       START
```

```
        title     "(MP64_3.ASM) multiword data prop test for final stage PEs"

        include   'OP_SYS.INC'    ; incl operating system equates & macros

N       equ       64
        org       x:$0
buff    ds        N               ; set aside buffer space for N pts (real)
        org       y:$0
        ds        N               ; set aside buffer space for N pts (imag)

        org       p:$100
START
        init_PE                   ; init this PE
        move      #-1,m0          ; linear access
        move      #1,n0           ; w/ offset inc of 1
;       move      #0,m0           ; bit-reverse access
;       move      #(N/2),n0       ; w/ offset inc associated w/ bit-rev access

LP      move      #buff,r0        ; initialize input data ptr register
        do        #N,lp1          ; transfer each of the input values
        nop                       ; from previous input/PE.  First loop
        get_PE    d0.s            ; for all of the real parts of the results,
        move      d0.s,x:(r0)+n0
        nop                       ; (NOPs for DO instruction restrictions)
lp1
        move      #buff,r0        ; initialize input data ptr register
        do        #N,lp2          ; then loop for all of the imag parts
        nop                       ; (both with normal output ordering)
        get_PE    d0.s
        move      d0.s,y:(r0)+n0
        nop                       ; (NOPs for DO instruction restrictions)
lp2
                                  ; <INSERT DATA PROCESSING CODE HERE>

        move      #buff,r0        ; initialize output data ptr register
        do        #N,lp3          ; transfer each of the output results
        put_OUT   x:(r0)+n0       ; to next PE in the array.  First loop
lp3                               ; for all of the real parts of the results,

        move      #buff,r0        ; initialize output data ptr register
        do        #N,lp4          ; then loop for all of the imag parts
        put_OUT   y:(r0)+n0       ; (both with normal output ordering)
lp4
        jmp       LP              ; continue infinite loop
        stop
        end       START
```

```
        title     "(MP128_1.ASM) multiword data prop test for initial stage PEs"

        include   'OP_SYS.INC'    ; incl operating system equates & macros

N       equ       128
        org       x:$0
buff    ds        N               ; set aside buffer space for N pts (real)
        org       y:$0
        ds        N               ; set aside buffer space for N pts (imag)

        org       p:$100
START
        init_PE                   ; init this PE
        move      #-1,m0          ; linear access
        move      #1,n0           ; w/ offset inc of 1
;       move      #0,m0           ; bit-reverse access
;       move      #(N/2),n0       ; w/ offset inc associated w/ bit-rev access

LP      move      #buff,r0        ; initialize input data ptr register
        do        #N,lp1          ; transfer each of the input values
        get_IN    d0.s            ; from previous input/PE.  First loop
        move      d0.s,x:(r0)+n0  ; for all of the real parts of the results,
lp1
        move      #buff,r0        ; initialize input data ptr register
        do        #N,lp2          ; then loop for all of the imag parts
        get_IN    d0.s            ; (both with normal output ordering)
        move      d0.s,y:(r0)+n0
lp2
                                  ; <INSERT DATA PROCESSING CODE HERE>

        move      #buff,r0        ; initialize output data ptr register
        do        #N,lp3          ; transfer each of the output results
        nop                       ; to next PE in the array.  First loop
        put_PE    x:(r0)+n0       ; for all of the real parts of the results,
        nop                       ; (NOPs for DO instruction restrictions)
lp3
        move      #buff,r0        ; initialize output data ptr register
        do        #N,lp4          ; then loop for all of the imag parts
        nop                       ; (both with normal output ordering)
        put_PE    y:(r0)+n0
        nop                       ; (NOPs for DO instruction restrictions)
lp4
        jmp       LP              ; continue infinite loop
        stop
        end       START
```

```
        title     "(MP128_2.ASM) multiword data prop test for intermed stage PEs"

        include   'OP_SYS.INC'     ; incl operating system equates & macros

N       equ       128
        org       x:$0
buff    ds        N                ; set aside buffer space for N pts (real)
        org       y:$0
        ds        N                ; set aside buffer space for N pts (imag)

        org       p:$100
START
        init_PE                    ; init this PE
        move      #-1,m0           ; linear access
        move      #1,n0            ; w/ offset inc of 1
;       move      #0,m0            ; bit-reverse access
;       move      #(N/2),n0        ; w/ offset inc associated w/ bit-rev access

LP      move      #buff,r0         ; initialize input data ptr register
        do        #N,lp1           ; transfer each of the input values
        nop                        ; from previous input/PE.  First loop
        get_PE    d0.s             ; for all of the real parts of the results,
        move      d0.s,x:(r0)+n0
        nop                        ; (NOPs for DO instruction restrictions)
lp1
        move      #buff,r0         ; initialize input data ptr register
        do        #N,lp2           ; then loop for all of the imag parts
        nop                        ; (both with normal output ordering)
        get_PE    d0.s
        move      d0.s,y:(r0)+n0
        nop                        ; (NOPs for DO instruction restrictions)
lp2
                                   ; <INSERT DATA PROCESSING CODE HERE>

        move      #buff,r0         ; initialize output data ptr register
        do        #N,lp3           ; transfer each of the output results
        nop                        ; to next PE in the array.  First loop
        put_PE    x:(r0)+n0        ; for all of the real parts of the results,
        nop                        ; (NOPs for DO instruction restrictions)
lp3
        move      #buff,r0         ; initialize output data ptr register
        do        #N,lp4           ; then loop for all of the imag parts
        nop                        ; (both with normal output ordering)
        put_PE    y:(r0)+n0
        nop                        ; (NOPs for DO instruction restrictions)
lp4
        jmp       LP               ; continue infinite loop
        stop
        end       START
```

```
        title      "(MP128_3.ASM) multiword data prop test for final stage PEs"

        include    'OP_SYS.INC'      ; incl operating system equates & macros

N       equ        128
        org        x:$0
buff    ds         N                 ; set aside buffer space for N pts (real)
        org        y:$0
        ds         N                 ; set aside buffer space for N pts (imag)

        org        p:$100
START
        init_PE                      ; init this PE
        move       #-1,m0            ; linear access
        move       #1,n0             ; w/ offset inc of 1
;       move       #0,m0             ; bit-reverse access
;       move       #(N/2),n0         ; w/ offset inc associated w/ bit-rev access

LP      move       #buff,r0          ; initialize input data ptr register
        do         #N,lp1            ; transfer each of the input values
        nop                          ; from previous input/PE. First loop
        get_PE     d0.s              ; for all of the real parts of the results,
        move       d0.s,x:(r0)+n0
        nop                          ; (NOPs for DO instruction restrictions)
lp1
        move       #buff,r0          ; initialize input data ptr register
        do         #N,lp2            ; then loop for all of the imag parts
        nop                          ; (both with normal output ordering)
        get_PE     d0.s
        move       d0.s,y:(r0)+n0
        nop                          ; (NOPs for DO instruction restrictions)
lp2
                                     ; <INSERT DATA PROCESSING CODE HERE>

        move       #buff,r0          ; initialize output data ptr register
        do         #N,lp3            ; transfer each of the output results
        put_OUT    x:(r0)+n0         ; to next PE in the array. First loop
lp3                                  ; for all of the real parts of the results,

        move       #buff,r0          ; initialize output data ptr register
        do         #N,lp4            ; then loop for all of the imag parts
        put_OUT    y:(r0)+n0         ; (both with normal output ordering)
lp4
        jmp        LP                ; continue infinite loop
        stop
        end        START
```

```
        title     "(MP256_1.ASM) multiword data prop test for initial stage PEs"

        include   'OP_SYS.INC'     ; incl operating system equates & macros

N       equ       256
        org       x:$0
buff    ds        N                ; set aside buffer space for N pts (real)
        org       y:$0
        ds        N                ; set aside buffer space for N pts (imag)

        org       p:$100
START
        init_PE                    ; init this PE
        move      #-1,m0           ; linear access
        move      #1,n0            ; w/ offset inc of 1
;       move      #0,m0            ; bit-reverse access
;       move      #(N/2),n0        ; w/ offset inc associated w/ bit-rev access

LP      move      #buff,r0         ; initialize input data ptr register
        do        #N,lp1           ; transfer each of the input values
        get_IN    d0.s             ; from previous input/PE.  First loop
        move      d0.s,x:(r0)+n0   ; for all of the real parts of the results,
lp1
        move      #buff,r0         ; initialize input data ptr register
        do        #N,lp2           ; then loop for all of the imag parts
        get_IN    d0.s             ; (both with normal output ordering)
        move      d0.s,y:(r0)+n0
lp2
                                   ; <INSERT DATA PROCESSING CODE HERE>

        move      #buff,r0         ; initialize output data ptr register
        do        #N,lp3           ; transfer each of the output results
        nop                        ; to next PE in the array.  First loop
        put_PE    x:(r0)+n0        ; for all of the real parts of the results,
        nop                        ; (NOPs for DO instruction restrictions)
lp3
        move      #buff,r0         ; initialize output data ptr register
        do        #N,lp4           ; then loop for all of the imag parts
        nop                        ; (both with normal output ordering)
        put_PE    y:(r0)+n0
        nop                        ; (NOPs for DO instruction restrictions)
lp4
        jmp       LP               ; continue infinite loop
        stop
        end       START
```

```
        title     "(MP256_2.ASM) multiword data prop test for intermed stage PEs"

        include   'OP_SYS.INC'      ; incl operating system equates & macros

N       equ       256
        org       x:$0
buff    ds        N                 ; set aside buffer space for N pts (real)
        org       y:$0
        ds        N                 ; set aside buffer space for N pts (imag)

        org       p:$100
START
        init_PE                     ; init this PE
        move      #-1,m0            ; linear access
        move      #1,n0             ; w/ offset inc of 1
;       move      #0,m0             ; bit-reverse access
;       move      #(N/2),n0         ; w/ offset inc associated w/ bit-rev access

LP      move      #buff,r0          ; initialize input data ptr register
        do        #N,lp1            ; transfer each of the input values
        nop                         ; from previous input/PE.  First loop
        get_PE    d0.s              ; for all of the real parts of the results,
        move      d0.s,x:(r0)+n0
        nop                         ; (NOPs for DO instruction restrictions)
lp1
        move      #buff,r0          ; initialize input data ptr register
        do        #N,lp2            ; then loop for all of the imag parts
        nop                         ; (both with normal output ordering)
        get_PE    d0.s
        move      d0.s,y:(r0)+n0
        nop                         ; (NOPs for DO instruction restrictions)
lp2
                                    ; <INSERT DATA PROCESSING CODE HERE>

        move      #buff,r0          ; initialize output data ptr register
        do        #N,lp3            ; transfer each of the output results
        nop                         ; to next PE in the array.  First loop
        put_PE    x:(r0)+n0         ; for all of the real parts of the results,
        nop                         ; (NOPs for DO instruction restrictions)
lp3
        move      #buff,r0          ; initialize output data ptr register
        do        #N,lp4            ; then loop for all of the imag parts
        nop                         ; (both with normal output ordering)
        put_PE    y:(r0)+n0
        nop                         ; (NOPs for DO instruction restrictions)
lp4
        jmp       LP                ; continue infinite loop
        stop
        end       START
```

```
        title     "(MP256_3.ASM) multiword data prop test for final stage PEs"

        include   'OP_SYS.INC'      ; incl operating system equates & macros

N       equ       256
        org       x:$0
buff    ds        N                 ; set aside buffer space for N pts (real)
        org       y:$0
        ds        N                 ; set aside buffer space for N pts (imag)

        org       p:$100
START
        init_PE                     ; init this PE
        move      #-1,m0            ; linear access
        move      #1,n0             ; w/ offset inc of 1
;       move      #0,m0             ; bit-reverse access
;       move      #(N/2),n0         ; w/ offset inc associated w/ bit-rev access

LP      move      #buff,r0          ; initialize input data ptr register
        do        #N,lp1            ; transfer each of the input values
        nop                         ; from previous input/PE. First loop
        get_PE    d0.s              ; for all of the real parts of the results,
        move      d0.s,x:(r0)+n0
        nop                         ; (NOPs for DO instruction restrictions)
lp1
        move      #buff,r0          ; initialize input data ptr register
        do        #N,lp2            ; then loop for all of the imag parts
        nop                         ; (both with normal output ordering)
        get_PE    d0.s
        move      d0.s,y:(r0)+n0
        nop                         ; (NOPs for DO instruction restrictions)
lp2
                                    ; <INSERT DATA PROCESSING CODE HERE>

        move      #buff,r0          ; initialize output data ptr register
        do        #N,lp3            ; transfer each of the output results
        put_OUT   x:(r0)+n0         ; to next PE in the array. First loop
lp3                                 ; for all of the real parts of the results,

        move      #buff,r0          ; initialize output data ptr register
        do        #N,lp4            ; then loop for all of the imag parts
        put_OUT   y:(r0)+n0         ; (both with normal output ordering)
lp4
        jmp       LP                ; continue infinite loop
        stop
        end       START
```

```
        title      "(MP512_1.ASM) multiword data prop test for initial stage PEs"

        include    'OP_SYS.INC'    ; incl operating system equates & macros

N       equ        512
        org        x:$0
buff    ds         N               ; set aside buffer space for N pts (real)
        org        y:$0
        ds         N               ; set aside buffer space for N pts (imag)

        org        p:$100
START
        init_PE                    ; init this PE
        move       #-1,m0          ; linear access
        move       #1,n0           ; w/ offset inc of 1
;       move       #0,m0           ; bit-reverse access
;       move       #(N/2),n0       ; w/ offset inc associated w/ bit-rev access

LP      move       #buff,r0        ; initialize input data ptr register
        do         #N,lp1          ; transfer each of the input values
        get_IN     d0.s            ; from previous input/PE. First loop
        move       d0.s,x:(r0)+n0  ; for all of the real parts of the results,
lp1
        move       #buff,r0        ; initialize input data ptr register
        do         #N,lp2          ; then loop for all of the imag parts
        get_IN     d0.s            ; (both with normal output ordering)
        move       d0.s,y:(r0)+n0
lp2
                                   ; <INSERT DATA PROCESSING CODE HERE>

        move       #buff,r0        ; initialize output data ptr register
        do         #N,lp3          ; transfer each of the output results
        nop                        ; to next PE in the array. First loop
        put_PE     x:(r0)+n0       ; for all of the real parts of the results,
        nop                        ; (NOPs for DO instruction restrictions)
lp3
        move       #buff,r0        ; initialize output data ptr register
        do         #N,lp4          ; then loop for all of the imag parts
        nop                        ; (both with normal output ordering)
        put_PE     y:(r0)+n0
        nop                        ; (NOPs for DO instruction restrictions)
lp4
        jmp        LP              ; continue infinite loop
        stop
        end        START
```

```
        title     "(MP512_2.ASM) multiword data prop test for intermed stage PEs"

        include   'OP_SYS.INC'    ; incl operating system equates & macros

N       equ       512
        org       x:$0
buff    ds        N               ; set aside buffer space for N pts (real)
        org       y:$0
        ds        N               ; set aside buffer space for N pts (imag)

        org       p:$100
START
        init_PE                   ; init this PE
        move      #-1,m0          ; linear access
        move      #1,n0           ; w/ offset inc of 1
;       move      #0,m0           ; bit-reverse access
;       move      #(N/2),n0       ; w/ offset inc associated w/ bit-rev access

LP      move      #buff,r0        ; initialize input data ptr register
        do        #N,lp1          ; transfer each of the input values
        nop                       ; from previous input/PE.  First loop
        get_PE    d0.s            ; for all of the real parts of the results,
        move      d0.s,x:(r0)+n0
        nop                       ; (NOPs for DO instruction restrictions)
lp1
        move      #buff,r0        ; initialize input data ptr register
        do        #N,lp2          ; then loop for all of the imag parts
        nop                       ; (both with normal output ordering)
        get_PE    d0.s
        move      d0.s,y:(r0)+n0
        nop                       ; (NOPs for DO instruction restrictions)
lp2
                                  ; <INSERT DATA PROCESSING CODE HERE>

        move      #buff,r0        ; initialize output data ptr register
        do        #N,lp3          ; transfer each of the output results
        nop                       ; to next PE in the array.  First loop
        put_PE    x:(r0)+n0       ; for all of the real parts of the results,
        nop                       ; (NOPs for DO instruction restrictions)
lp3
        move      #buff,r0        ; initialize output data ptr register
        do        #N,lp4          ; then loop for all of the imag parts
        nop                       ; (both with normal output ordering)
        put_PE    y:(r0)+n0
        nop                       ; (NOPs for DO instruction restrictions)
lp4
        jmp       LP              ; continue infinite loop
        stop
        end       START
```

```
        title     "(MP512_3.ASM) multiword data prop test for final stage PEs"

        include   'OP_SYS.INC'    ; incl operating system equates & macros

N       equ       512
        org       x:$0
buff    ds        N               ; set aside buffer space for N pts (real)
        org       y:$0
        ds        N               ; set aside buffer space for N pts (imag)

        org       p:$100
START
        init_PE                   ; init this PE
        move      #-1,m0          ; linear access
        move      #1,n0           ; w/ offset inc of 1
;       move      #0,m0           ; bit-reverse access
;       move      #(N/2),n0       ; w/ offset inc associated w/ bit-rev access

LP      move      #buff,r0        ; initialize input data ptr register
        do        #N,lp1          ; transfer each of the input values
        nop                       ; from previous input/PE.  First loop
        get_PE    d0.s            ; for all of the real parts of the results,
        move      d0.s,x:(r0)+n0
        nop                       ; (NOPs for DO instruction restrictions)
lp1
        move      #buff,r0        ; initialize input data ptr register
        do        #N,lp2          ; then loop for all of the imag parts
        nop                       ; (both with normal output ordering)
        get_PE    d0.s
        move      d0.s,y:(r0)+n0
        nop                       ; (NOPs for DO instruction restrictions)
lp2
                                  ; <INSERT DATA PROCESSING CODE HERE>

        move      #buff,r0        ; initialize output data ptr register
        do        #N,lp3          ; transfer each of the output results
        put_OUT   x:(r0)+n0       ; to next PE in the array.  First loop
lp3                               ; for all of the real parts of the results,

        move      #buff,r0        ; initialize output data ptr register
        do        #N,lp4          ; then loop for all of the imag parts
        put_OUT   y:(r0)+n0       ; (both with normal output ordering)
lp4
        jmp       LP              ; continue infinite loop
        stop
        end       START
```

```
        title     "(DF1_1.ASM) digital filter test for 1st stage PEs (B=1)"

        include   'OP_SYS.INC'       ; include operating system equates & macros

B       equ       1                  ; B cascaded real biquad IIR filters

        org       x:$100
        buffer    M,2                ; modulo-access 2-element delayed data array:
DATA    dc        0.0                ; w1(n-2) delayed data element w/ IC
        dc        0.0                ; w1(n-1) delayed data element w/ IC
        endbuf

        org       y:$100
        buffer    M,4                ; modulo-access 4-element coeff array:
COEFF   dc        0.10               ; a21 coeff (1st set)
        dc        0.20               ; a11 coeff
        dc        0.30               ; b21 coeff
        dc        0.40               ; b11 coeff
        endbuf

        org       p:$100
START
        init_PE                      ; initialize this PE
        init_DF   B,DATA,COEFF       ; initialize the digital filter operation

LP      get_IN    d0.s               ; get system input x(n)
        calc_DF   B,d0.s             ; calculate the next output of the dig filter
        put_PE    d0.s               ; transfer y(n) to next PE in array

        JMP       LP                 ; repeat to form an infinite loop
        STOP
```

```
        title     "(DF1_2.ASM) digital filter test for stage 2 PEs (B=1)"

        include   'OP_SYS.INC'     ; include operating system equates & macros

B       equ       1                ; B cascaded real biquad IIR filters

        org       x:$100
        buffer    M,2              ; modulo-access 2-element delayed data array:
DATA    dc        0.0              ; w1(n-2) delayed data element w/ IC
        dc        0.0              ; w1(n-1) delayed data element w/ IC
        endbuf

        org       y:$100
        buffer    M,4              ; modulo-access 4-element coeff array:
COEFF   dc        0.10             ; a21 coeff (1st set)
        dc        0.20             ; a11 coeff
        dc        0.30             ; b21 coeff
        dc        0.40             ; b11 coeff
        endbuf

        org       p:$100
START
        init_PE                    ; initialize this PE
        init_DF   B,DATA,COEFF     ; initialize the digital filter operation

LP      get_PE    D0.S             ; get input x(n) from previous PE in array
        calc_DF   B,d0.s           ; calculate the next output of the dig filter
        put_PE    d0.s             ; transfer y(n) to next PE in array

        JMP       LP               ; repeat to form an infinite loop
        STOP
```

```
        title      "(DF1_3.ASM) digital filter test for 3rd/final PEs (B=1)"

        include    'OP_SYS.INC'     ; include operating system equates & macros

B       equ        1                ; B cascaded real biquad IIR filters

        org        x:$100
        buffer     M,2              ; modulo-access 2-element delayed data array:
DATA    dc         0.0              ; w1(n-2) delayed data element w/ IC
        dc         0.0              ; w1(n-1) delayed data element w/ IC
        endbuf

        org        y:$100
        buffer     M,4              ; modulo-access 4-element coeff array:
COEFF   dc         0.10             ; a21 coeff (1st set)
        dc         0.20             ; a11 coeff
        dc         0.30             ; b21 coeff
        dc         0.40             ; b11 coeff
        endbuf

        org        p:$100
START
        init_PE                     ; initialize this PE
        init_DF    B,DATA,COEFF     ; initialize the digital filter operation

LP      get_PE     D0.S             ; get input x(n) from previous PE in array
        calc_DF    B,d0.s           ; calculate the next output of the dig filter
        put_OUT    D0.S             ; transfer y(n) to the output latch

        JMP        LP               ; repeat to form an infinite loop
        STOP
```

```
        title     "(DF2_1.ASM) digital filter test for 1st stage PEs (B=2)"

        include   'OP_SYS.INC'      ; include operating system equates & macros

B       equ       2                 ; B cascaded real biquad IIR filters

        org       x:$100
        buffer    M,4               ; modulo-access 4-element delayed data array:
DATA    dc        0.0               ; w1(n-2) delayed data element w/ IC
        dc        0.0               ; w1(n-1) delayed data element w/ IC
        dc        0.0               ; w2(n-2) delayed data element w/ IC
        dc        0.0               ; w2(n-1) delayed data element w/ IC
        endbuf

        org       y:$100
        buffer    M,8               ; modulo-access 8-element coeff array:
COEFF   dc        0.10              ; a21 coeff (1st set)
        dc        0.20              ; a11 coeff
        dc        0.30              ; b21 coeff
        dc        0.40              ; b11 coeff
        dc        0.50              ; a22 coeff (2nd set)
        dc        0.60              ; a12 coeff
        dc        0.70              ; b22 coeff
        dc        0.80              ; b12 coeff
        endbuf

        org       p:$100
START
        init_PE                     ; initialize this PE
        init_DF   B,DATA,COEFF      ; initialize the digital filter operation

LP      get_IN    d0.s              ; get system input x(n)
        calc_DF   B,d0.s            ; calculate the next output of the dig filter
        put_PE    d0.s              ; transfer y(n) to next PE in array

        JMP       LP                ; repeat to form an infinite loop
        STOP
```

```
        title     "(DF2_2.ASM) digital filter test for stage 2 PEs (B=2)"

        include   'OP_SYS.INC'     ; include operating system equates & macros

B       equ       2                ; B cascaded real biquad IIR filters

        org       x:$100
        buffer    M,4              ; modulo-access 4-element delayed data array:
DATA    dc        0.0              ; w1(n-2) delayed data element w/ IC
        dc        0.0              ; w1(n-1) delayed data element w/ IC
        dc        0.0              ; w2(n-2) delayed data element w/ IC
        dc        0.0              ; w2(n-1) delayed data element w/ IC
        endbuf

        org       y:$100
        buffer    M,8              ; modulo-access 8-element coeff array:
COEFF   dc        0.10             ; a21 coeff (1st set)
        dc        0.20             ; a11 coeff
        dc        0.30             ; b21 coeff
        dc        0.40             ; b11 coeff
        dc        0.50             ; a22 coeff (2nd set)
        dc        0.60             ; a12 coeff
        dc        0.70             ; b22 coeff
        dc        0.80             ; b12 coeff
        endbuf

        org       p:$100
START
        init_PE                    ; initialize this PE
        init_DF   B,DATA,COEFF     ; initialize the digital filter operation

LP      get_PE    D0.S             ; get input x(n) from previous PE in array
        calc_DF   B,d0.s           ; calculate the next output of the dig filter
        put_PE    d0.s             ; transfer y(n) to next PE in array

        JMP       LP               ; repeat to form an infinite loop
        STOP
```

```
        title     "(DF2_3.ASM) digital filter test for 3rd/final PEs (B=2)"

        include   'OP_SYS.INC'     ; include operating system equates & macros

B       equ       2                ; B cascaded real biquad IIR filters

        org       x:$100
        buffer    M,4              ; modulo-access 4-element delayed data array:
DATA    dc        0.0              ; w1(n-2) delayed data element w/ IC
        dc        0.0              ; w1(n-1) delayed data element w/ IC
        dc        0.0              ; w2(n-2) delayed data element w/ IC
        dc        0.0              ; w2(n-1) delayed data element w/ IC
        endbuf

        org       y:$100
        buffer    M,8              ; modulo-access 8-element coeff array:
COEFF   dc        0.10             ; a21 coeff (1st set)
        dc        0.20             ; a11 coeff
        dc        0.30             ; b21 coeff
        dc        0.40             ; b11 coeff
        dc        0.50             ; a22 coeff (2nd set)
        dc        0.60             ; a12 coeff
        dc        0.70             ; b22 coeff
        dc        0.80             ; b12 coeff
        endbuf

        org       p:$100
START
        init_PE                    ; initialize this PE
        init_DF   B,DATA,COEFF     ; initialize the digital filter operation

LP      get_PE    D0.S             ; get input x(n) from previous PE in array
        calc_DF   B,d0.s           ; calculate the next output of the dig filter
        put_OUT   D0.S             ; transfer y(n) to the output latch

        JMP       LP               ; repeat to form an infinite loop
        STOP
```

```
        title     "(DF3_1.ASM) digital filter test for 1st stage PEs (B=3)"

        include   'OP_SYS.INC'     ; include operating system equates & macros

B       equ       3                ; B cascaded real biquad IIR filters

        org       x:$100
        buffer    M,6              ; modulo-access 6-element delayed data array:
DATA    dc        0.0              ; w1(n-2) delayed data element w/ IC
        dc        0.0              ; w1(n-1) delayed data element w/ IC
        dc        0.0              ; w2(n-2) delayed data element w/ IC
        dc        0.0              ; w2(n-1) delayed data element w/ IC
        dc        0.0              ; w3(n-2) delayed data element w/ IC
        dc        0.0              ; w3(n-1) delayed data element w/ IC
        endbuf

        org       y:$100
        buffer    M,12             ; modulo-access 12-element coeff array:
COEFF   dc        0.10             ; a21 coeff (1st set)
        dc        0.20             ; a11 coeff
        dc        0.30             ; b21 coeff
        dc        0.40             ; b11 coeff
        dc        0.50             ; a22 coeff (2nd set)
        dc        0.60             ; a12 coeff
        dc        0.70             ; b22 coeff
        dc        0.80             ; b12 coeff
        dc        0.90             ; a23 coeff (3rd set)
        dc        1.00             ; a13 coeff
        dc        1.10             ; b23 coeff
        dc        1.20             ; b13 coeff
        endbuf

        org       p:$100
START
        init_PE                    ; initialize this PE
        init_DF   B,DATA,COEFF     ; initialize the digital filter operation

LP      get_IN    d0.s             ; get system input x(n)
        calc_DF   B,d0.s           ; calculate the next output of the dig filter
        put_PE    d0.s             ; transfer y(n) to next PE in array

        JMP       LP               ; repeat to form an infinite loop
        STOP
```

```
        title       "(DF3_2.ASM) digital filter test for stage 2 PEs (B=3)"

        include     'OP_SYS.INC'    ; include operating system equates & macros

B       equ         3               ; B cascaded real biquad IIR filters

        org         x:$100
        buffer      M,6             ; modulo-access 6-element delayed data array:
DATA    dc          0.0             ; w1(n-2) delayed data element w/ IC
        dc          0.0             ; w1(n-1) delayed data element w/ IC
        dc          0.0             ; w2(n-2) delayed data element w/ IC
        dc          0.0             ; w2(n-1) delayed data element w/ IC
        dc          0.0             ; w3(n-2) delayed data element w/ IC
        dc          0.0             ; w3(n-1) delayed data element w/ IC
        endbuf

        org         y:$100
        buffer      M,12            ; modulo-access 12-element coeff array:
COEFF   dc          0.10            ; a21 coeff (1st set)
        dc          0.20            ; a11 coeff
        dc          0.30            ; b21 coeff
        dc          0.40            ; b11 coeff
        dc          0.50            ; a22 coeff (2nd set)
        dc          0.60            ; a12 coeff
        dc          0.70            ; b22 coeff
        dc          0.80            ; b12 coeff
        dc          0.90            ; a23 coeff (3rd set)
        dc          1.00            ; a13 coeff
        dc          1.10            ; b23 coeff
        dc          1.20            ; b13 coeff
        endbuf

        org         p:$100
START
        init_PE                     ; initialize this PE
        init_DF     B,DATA,COEFF    ; initialize the digital filter operation

LP      get_PE      D0.S            ; get input x(n) from previous PE in array
        calc_DF     B,d0.s          ; calculate the next output of the dig filter
        put_PE      d0.s            ; transfer y(n) to next PE in array

        JMP         LP              ; repeat to form an infinite loop
        STOP
```

```
        title     "(DF3_3.ASM) digital filter test for 3rd/final PEs (B=3)"

        include   'OP_SYS.INC'     ; include operating system equates & macros

B       equ       3                ; B cascaded real biquad IIR filters

        org       x:$100
        buffer    M,6              ; modulo-access 6-element delayed data array:
DATA    dc        0.0              ; w1(n-2) delayed data element w/ IC
        dc        0.0              ; w1(n-1) delayed data element w/ IC
        dc        0.0              ; w2(n-2) delayed data element w/ IC
        dc        0.0              ; w2(n-1) delayed data element w/ IC
        dc        0.0              ; w3(n-2) delayed data element w/ IC
        dc        0.0              ; w3(n-1) delayed data element w/ IC
        endbuf

        org       y:$100
        buffer    M,12             ; modulo-access 12-element coeff array:
COEFF   dc        0.10             ; a21 coeff (1st set)
        dc        0.20             ; a11 coeff
        dc        0.30             ; b21 coeff
        dc        0.40             ; b11 coeff
        dc        0.50             ; a22 coeff (2nd set)
        dc        0.60             ; a12 coeff
        dc        0.70             ; b22 coeff
        dc        0.80             ; b12 coeff
        dc        0.90             ; a23 coeff (3rd set)
        dc        1.00             ; a13 coeff
        dc        1.10             ; b23 coeff
        dc        1.20             ; b13 coeff
        endbuf

        org       p:$100
START
        init_PE                    ; initialize this PE
        init_DF   B,DATA,COEFF     ; initialize the digital filter operation

LP      get_PE    D0.S             ; get input x(n) from previous PE in array
        calc_DF   B,d0.s           ; calculate the next output of the dig filter
        put_OUT   D0.S             ; transfer y(n) to the output latch

        JMP       LP               ; repeat to form an infinite loop
        STOP
```

```
        title     "(DF4_1.ASM) digital filter test for 1st stage PEs (B=4)"

        include   'OP_SYS.INC'      ; include operating system equates & macros

B       equ       4                 ; B cascaded real biquad IIR filters

        org       x:$100
        buffer    M,8               ; modulo-access 8-element delayed data array:
DATA    dc        0.0               ; w1(n-2) delayed data element w/ IC
        dc        0.0               ; w1(n-1) delayed data element w/ IC
        dc        0.0               ; w2(n-2) delayed data element w/ IC
        dc        0.0               ; w2(n-1) delayed data element w/ IC
        dc        0.0               ; w3(n-2) delayed data element w/ IC
        dc        0.0               ; w3(n-1) delayed data element w/ IC
        dc        0.0               ; w4(n-2) delayed data element w/ IC
        dc        0.0               ; w4(n-1) delayed data element w/ IC
        endbuf

        org       y:$100
        buffer    M,16              ; modulo-access 16-element coeff array:
COEFF   dc        0.10              ; a21 coeff (1st set)
        dc        0.20              ; a11 coeff
        dc        0.30              ; b21 coeff
        dc        0.40              ; b11 coeff
        dc        0.50              ; a22 coeff (2nd set)
        dc        0.60              ; a12 coeff
        dc        0.70              ; b22 coeff
        dc        0.80              ; b12 coeff
        dc        0.90              ; a23 coeff (3rd set)
        dc        1.00              ; a13 coeff
        dc        1.10              ; b23 coeff
        dc        1.20              ; b13 coeff
        dc        1.30              ; a24 coeff (4th set)
        dc        1.40              ; a14 coeff
        dc        1.50              ; b24 coeff
        dc        1.60              ; b14 coeff
        endbuf

        org       p:$100
START
        init_PE                     ; initialize this PE
        init_DF   B,DATA,COEFF      ; initialize the digital filter operation

LP      get_IN    d0.s              ; get system input x(n)
        calc_DF   B,d0.s            ; calculate the next output of the dig filter
        put_PE    d0.s              ; transfer y(n) to next PE in array

        JMP       LP                ; repeat to form an infinite loop
        STOP
```

```
        title     "(DF4_2.ASM) digital filter test for stage 2 PEs (B=4)"

        include   'OP_SYS.INC'    ; include operating system equates & macros

B       equ       4               ; B cascaded real biquad IIR filters

        org       x:$100
        buffer    M,8             ; modulo-access 8-element delayed data array:
DATA    dc        0.0             ; w1(n-2) delayed data element w/ IC
        dc        0.0             ; w1(n-1) delayed data element w/ IC
        dc        0.0             ; w2(n-2) delayed data element w/ IC
        dc        0.0             ; w2(n-1) delayed data element w/ IC
        dc        0.0             ; w3(n-2) delayed data element w/ IC
        dc        0.0             ; w3(n-1) delayed data element w/ IC
        dc        0.0             ; w4(n-2) delayed data element w/ IC
        dc        0.0             ; w4(n-1) delayed data element w/ IC
        endbuf

        org       y:$100
        buffer    M,16            ; modulo-access 16-element coeff array:
COEFF   dc        0.10            ; a21 coeff (1st set)
        dc        0.20            ; a11 coeff
        dc        0.30            ; b21 coeff
        dc        0.40            ; b11 coeff
        dc        0.50            ; a22 coeff (2nd set)
        dc        0.60            ; a12 coeff
        dc        0.70            ; b22 coeff
        dc        0.80            ; b12 coeff
        dc        0.90            ; a23 coeff (3rd set)
        dc        1.00            ; a13 coeff
        dc        1.10            ; b23 coeff
        dc        1.20            ; b13 coeff
        dc        1.30            ; a24 coeff (4th set)
        dc        1.40            ; a14 coeff
        dc        1.50            ; b24 coeff
        dc        1.60            ; b14 coeff
        endbuf

        org       p:$100
START
        init_PE                   ; initialize this PE
        init_DF   B,DATA,COEFF    ; initialize the digital filter operation

LP      get_PE    D0.S            ; get input x(n) from previous PE in array
        calc_DF   B,d0.s          ; calculate the next output of the dig filter
        put_PE    d0.s            ; transfer y(n) to next PE in array

        JMP       LP              ; repeat to form an infinite loop
        STOP
```

```
        title     "(DF4_3.ASM) digital filter test for 3rd/final PEs (B=4)"

        include   'OP_SYS.INC'     ; include operating system equates & macros

B       equ       4                ; B cascaded real biquad IIR filters

        org       x:$100
        buffer    M,8              ; modulo-access 8-element delayed data array:
DATA    dc        0.0              ; w1(n-2) delayed data element w/ IC
        dc        0.0              ; w1(n-1) delayed data element w/ IC
        dc        0.0              ; w2(n-2) delayed data element w/ IC
        dc        0.0              ; w2(n-1) delayed data element w/ IC
        dc        0.0              ; w3(n-2) delayed data element w/ IC
        dc        0.0              ; w3(n-1) delayed data element w/ IC
        dc        0.0              ; w4(n-2) delayed data element w/ IC
        dc        0.0              ; w4(n-1) delayed data element w/ IC
        endbuf

        org       y:$100
        buffer    M,16             ; modulo-access 16-element coeff array:
COEFF   dc        0.10             ; a21 coeff (1st set)
        dc        0.20             ; a11 coeff
        dc        0.30             ; b21 coeff
        dc        0.40             ; b11 coeff
        dc        0.50             ; a22 coeff (2nd set)
        dc        0.60             ; a12 coeff
        dc        0.70             ; b22 coeff
        dc        0.80             ; b12 coeff
        dc        0.90             ; a23 coeff (3rd set)
        dc        1.00             ; a13 coeff
        dc        1.10             ; b23 coeff
        dc        1.20             ; b13 coeff
        dc        1.30             ; a24 coeff (4th set)
        dc        1.40             ; a14 coeff
        dc        1.50             ; b24 coeff
        dc        1.60             ; b14 coeff
        endbuf

        org       p:$100
START
        init_PE                    ; initialize this PE
        init_DF   B,DATA,COEFF     ; initialize the digital filter operation

LP      get_PE    D0.S             ; get input x(n) from previous PE in array
        calc_DF   B,d0.s           ; calculate the next output of the dig filter
        put_OUT   D0.S             ; transfer y(n) to the output latch

        JMP       LP               ; repeat to form an infinite loop
        STOP
```

```
        title     "(FFT8_1.ASM) FFT test for initial stage PEs (N=8)"

        include   'OP_SYS.INC'     ; incl operating system equates & macros

N       equ       8
        org       x:$0
fftdat  ds        N                ; set aside buffer space for N pts (real)
        org       y:$0
        ds        N                ; set aside buffer space for N pts (imag)

        org       p:$100
START
        init_PE                    ; init this PE

LP      move      #-1,m0           ; linear access
        move      #1,n0            ; w/ offset inc of 1
;       move      #0,m0            ; bit-reverse access
;       move      #(N/2),n0        ; w/ offset inc associated w/ bit-rev access

        move      #fftdat,r0       ; initialize input data ptr register
        do        #N,lp1           ; transfer each of the input values
        get_IN    d0.s             ; from previous input/PE.  First loop
        move      d0.s,x:(r0)+n0   ; for all of the real parts of the results,
lp1
        move      #fftdat,r0       ; initialize input data ptr register
        do        #N,lp2           ; then loop for all of the imag parts
        get_IN    d0.s             ; (both with normal output ordering)
        move      d0.s,y:(r0)+n0
lp2
                                                 ; multipass FFT testing:
        mpfft     N,fftdat,ROM_TBL,ROM_LEN/2,1,1   ; perform N-pt FFT (pass 1)
;       mpfft     N,fftdat,ROM_TBL,ROM_LEN/2,2,1   ; perform N-pt FFT (pass 2)
;       mpfft     N,fftdat,ROM_TBL,ROM_LEN/2,3,1   ; perform N-pt FFT (pass 3)
                                                 ; (w/ half-cycle tables)

        move      #-1,m0           ; linear access
        move      #1,n0            ; w/ offset inc of 1
;       move      #0,m0            ; bit-reverse access
;       move      #(N/2),n0        ; w/ offset inc associated w/ bit-rev access

        move      #fftdat,r0       ; initialize output data ptr register
        do        #N,lp3           ; transfer each of the output results
        nop                        ; to next PE in the array.  First loop
        put_PE    x:(r0)+n0        ; for all of the real parts of the results,
        nop                        ; (NOPs for DO instruction restrictions)
lp3
        move      #fftdat,r0       ; initialize output data ptr register
        do        #N,lp4           ; then loop for all of the imag parts
        nop                        ; (both with normal output ordering)
```

```
        put_PE    y:(r0)+n0
        nop                          ; (NOPs for DO instruction restrictions)
lp4
        jmp       LP                 ; continue infinite loop
        stop
        end       START
```

```
        title      "(FFT8_2.ASM) FFT test for intermed stage PEs (N=8)"

        include    'OP_SYS.INC'     ; incl operating system equates & macros

N       equ        8
        org        x:$0
fftdat  ds         N                ; set aside buffer space for N pts (real)
        org        y:$0
        ds         N                ; set aside buffer space for N pts (imag)

        org        p:$100
START
        init_PE                     ; init this PE

LP      move       #-1,m0           ; linear access
        move       #1,n0            ; w/ offset inc of 1
;       move       #0,m0            ; bit-reverse access
;       move       #(N/2),n0        ; w/ offset inc associated w/ bit-rev access

        move       #fftdat,r0       ; initialize input data ptr register
        do         #N,lp1           ; transfer each of the input values
        nop                         ; from previous input/PE.  First loop
        get_PE     d0.s             ; for all of the real parts of the results,
        move       d0.s,x:(r0)+n0
        nop                         ; (NOPs for DO instruction restrictions)
lp1
        move       #fftdat,r0       ; initialize input data ptr register
        do         #N,lp2           ; then loop for all of the imag parts
        nop                         ; (both with normal output ordering)
        get_PE     d0.s
        move       d0.s,y:(r0)+n0
        nop                         ; (NOPs for DO instruction restrictions)
lp2
                                                      ; multipass FFT testing:
;       mpfft      N,fftdat,ROM_TBL,ROM_LEN/2,1,1    ; perform N-pt FFT (pass 1)
        mpfft      N,fftdat,ROM_TBL,ROM_LEN/2,2,1    ; perform N-pt FFT (pass 2)
;       mpfft      N,fftdat,ROM_TBL,ROM_LEN/2,3,1    ; perform N-pt FFT (pass 3)
                                                      ; (w/ half-cycle tables)

        move       #-1,m0           ; linear access
        move       #1,n0            ; w/ offset inc of 1
;       move       #0,m0            ; bit-reverse access
;       move       #(N/2),n0        ; w/ offset inc associated w/ bit-rev access

        move       #fftdat,r0       ; initialize output data ptr register
        do         #N,lp3           ; transfer each of the output results
        nop                         ; to next PE in the array.  First loop
        put_PE     x:(r0)+n0        ; for all of the real parts of the results,
        nop                         ; (NOPs for DO instruction restrictions)
```

```
lp3
        move      #fftdat,r0      ; initialize output data ptr register
        do        #N,lp4          ; then loop for all of the imag parts
        nop                       ; (both with normal output ordering)
        put_PE    y:(r0)+n0
        nop                       ; (NOPs for DO instruction restrictions)
lp4
        jmp       LP              ; continue infinite loop
        stop
        end       START
```

```
        title       "(FFT8_3.ASM) FFT test for final stage PEs (N=8)"

        include     'OP_SYS.INC'      ; incl operating system equates & macros

N       equ         8
        org         x:$0
fftdat  ds          N                 ; set aside buffer space for N pts (real)
        org         y:$0
        ds          N                 ; set aside buffer space for N pts (imag)

        org         p:$100
START
        init_PE                       ; init this PE

LP      move        #-1,m0            ; linear access
        move        #1,n0             ; w/ offset inc of 1
;       move        #0,m0             ; bit-reverse access
;       move        #(N/2),n0         ; w/ offset inc associated w/ bit-rev access

        move        #fftdat,r0        ; initialize input data ptr register
        do          #N,lp1            ; transfer each of the input values
        nop                           ; from previous input/PE.  First loop
        get_PE      d0.s              ; for all of the real parts of the results,
        move        d0.s,x:(r0)+n0
        nop                           ; (NOPs for DO instruction restrictions
lp1
        move        #fftdat,r0        ; initialize input data ptr register
        do          #N,lp2            ; then loop for all of the imag parts
        nop                           ; (both with normal output ordering)
        get_PE      d0.s
        move        d0.s,y:(r0)+n0
        nop                           ; (NOPs for DO instruction restrictions)
lp2
                                                  ; multipass FFT testing:
;       mpfft       N,fftdat,ROM_TBL,ROM_LEN/2,1,1    ; perform N-pt FFT (pass 1)
;       mpfft       N,fftdat,ROM_TBL,ROM_LEN/2,2,1    ; perform N-pt FFT (pass 2)
        mpfft       N,fftdat,ROM_TBL,ROM_LEN/2,3,1    ; perform N-pt FFT (pass 3)
                                                  ; (w/ half-cycle tables)

;       move        #-1,m0            ; linear access
;       move        #1,n0             ; w/ offset inc of 1
        move        #0,m0             ; bit-reverse access
        move        #(N/2),n0         ; w/ offset inc associated w/ bit-rev access

        move        #fftdat,r0        ; initialize output data ptr register
        do          #N,lp3            ; transfer each of the output results
        put_OUT     x:(r0)+n0         ; to next PE in the array.  First loop
lp3                                   ; for all of the real parts of the results,
```

```
        move     #fftdat,r0      ; initialize output data ptr register
        do       #N,lp4          ; then loop for all of the imag parts
        put_OUT  y:(r0)+n0       ; (both with normal output ordering)
lp4
        jmp      LP              ; continue infinite loop
        stop
        end      START
```

```
        title     "(FFT16_1.ASM) FFT test for initial stage PEs (N=16)"

        include   'OP_SYS.INC'      ; incl operating system equates & macros

N       equ       16
        org       x:$0
fftdat  ds        N                 ; set aside buffer space for N pts (real)
        org       y:$0
        ds        N                 ; set aside buffer space for N pts (imag)

        org       p:$100
START
        init_PE                     ; init this PE

LP      move      #-1,m0            ; linear access
        move      #1,n0             ; w/ offset inc of 1
;       move      #0,m0             ; bit-reverse access
;       move      #(N/2),n0         ; w/ offset inc associated w/ bit-rev access

        move      #fftdat,r0        ; initialize input data ptr register
        do        #N,lp1            ; transfer each of the input values
        get_IN    d0.s              ; from previous input/PE.  First loop
        move      d0.s,x:(r0)+n0    ; for all of the real parts of the results,
lp1
        move      #fftdat,r0        ; initialize input data ptr register
        do        #N,lp2            ; then loop for all of the imag parts
        get_IN    d0.s              ; (both with normal output ordering)
        move      d0.s,y:(r0)+n0
lp2
                                                ; multipass FFT testing:
        mpfft     N,fftdat,ROM_TBL,ROM_LEN/2,1,2    ; perform N-pt FFT (pass 1-2)
;       mpfft     N,fftdat,ROM_TBL,ROM_LEN/2,3,1    ; perform N-pt FFT (pass 3)
;       mpfft     N,fftdat,ROM_TBL,ROM_LEN/2,4,1    ; perform N-pt FFT (pass 4)
                                                ; (w/ half-cycle tables)

        move      #-1,m0            ; linear access
        move      #1,n0             ; w/ offset inc of 1
;       move      #0,m0             ; bit-reverse access
;       move      #(N/2),n0         ; w/ offset inc associated w/ bit-rev access

        move      #fftdat,r0        ; initialize output data ptr register
        do        #N,lp3            ; transfer each of the output results
        nop                         ; to next PE in the array.  First loop
        put_PE    x:(r0)+n0         ; for all of the real parts of the results,
        nop                         ; (NOPs for DO instruction restrictions)
lp3
        move      #fftdat,r0        ; initialize output data ptr register
        do        #N,lp4            ; then loop for all of the imag parts
        nop                         ; (both with normal output ordering)
```

```
        put_PE    y:(r0)+n0
        nop                         ; (NOPs for DO instruction restrictions)
lp4
        jmp       LP                ; continue infinite loop
        stop
        end       START
```

```
        title     "(FFT16_2.ASM) FFT test for intermed stage PEs (N=16)"

        include   'OP_SYS.INC'       ; incl operating system equates & macros

N       equ       16
        org       x:$0
fftdat  ds        N                  ; set aside buffer space for N pts (real)
        org       y:$0
        ds        N                  ; set aside buffer space for N pts (imag)

        org       p:$100
START
        init_PE                      ; init this PE

LP      move      #-1,m0             ; linear access
        move      #1,n0              ; w/ offset inc of 1
;       move      #0,m0              ; bit-reverse access
;       move      #(N/2),n0          ; w/ offset inc associated w/ bit-rev access

        move      #fftdat,r0         ; initialize input data ptr register
        do        #N,lp1             ; transfer each of the input values
        nop                          ; from previous input/PE.  First loop
        get_PE    d0.s               ; for all of the real parts of the results,
        move      d0.s,x:(r0)+n0
        nop                          ; (NOPs for DO instruction restrictions)
lp1
        move      #fftdat,r0         ; initialize input data ptr register
        do        #N,lp2             ; then loop for all of the imag parts
        nop                          ; (both with normal output ordering)
        get_PE    d0.s
        move      d0.s,y:(r0)+n0
        nop                          ; (NOPs for DO instruction restrictions)
lp2
                                               ; multipass FFT testing:
;       mpfft     N,fftdat,ROM_TBL,ROM_LEN/2,1,2   ; perform N-pt FFT (pass 1-2)
        mpfft     N,fftdat,ROM_TBL,ROM_LEN/2,3,1   ; perform N-pt FFT (pass 3)
;       mpfft     N,fftdat,ROM_TBL,ROM_LEN/2,4,1   ; perform N-pt FFT (pass 4)
                                               ; (w/ half-cycle tables)

        move      #-1,m0             ; linear access
        move      #1,n0              ; w/ offset inc of 1
;       move      #0,m0              ; bit-reverse access
;       move      #(N/2),n0          ; w/ offset inc associated w/ bit-rev access

        move      #fftdat,r0         ; initialize output data ptr register
        do        #N,lp3             ; transfer each of the output results
        nop                          ; to next PE in the array.  First loop
        put_PE    x:(r0)+n0          ; for all of the real parts of the results,
        nop                          ; (NOPs for DO instruction restrictions)
```

```
lp3
        move    #fftdat,r0      ; initialize output data ptr register
        do      #N,lp4          ; then loop for all of the imag parts
        nop                     ; (both with normal output ordering)
        put_PE  y:(r0)+n0
        nop                     ; (NOPs for DO instruction restrictions)
lp4
        jmp     LP              ; continue infinite loop
        stop
        end     START
```

```
        title     "(FFT16_3.ASM) FFT test for final stage PEs (N=16)"

        include   'OP_SYS.INC'      ; incl operating system equates & macros

N       equ       16
        org       x:$0
fftdat  ds        N                 ; set aside buffer space for N pts (real)
        org       y:$0
        ds        N                 ; set aside buffer space for N pts (imag)

        org       p:$100
START
        init_PE                     ; init this PE

LP      move      #-1,m0            ; linear access
        move      #1,n0             ; w/ offset inc of 1
;       move      #0,m0             ; bit-reverse access
;       move      #(N/2),n0         ; w/ offset inc associated w/ bit-rev access

        move      #fftdat,r0        ; initialize input data ptr register
        do        #N,lp1            ; transfer each of the input values
        nop                         ; from previous input/PE.  First loop
        get_PE    d0.s              ; for all of the real parts of the results,
        move      d0.s,x:(r0)+n0
        nop                         ; (NOPs for DO instruction restrictions
lp1
        move      #fftdat,r0        ; initialize input data ptr register
        do        #N,lp2            ; then loop for all of the imag parts
        nop                         ; (both with normal output ordering)
        get_PE    d0.s
        move      d0.s,y:(r0)+n0
        nop                         ; (NOPs for DO instruction restrictions)
lp2
                                            ; multipass FFT testing:
;       mpfft     N,fftdat,ROM_TBL,ROM_LEN/2,1,2    ; perform N-pt FFT (pass 1-2)
;       mpfft     N,fftdat,ROM_TBL,ROM_LEN/2,3,1    ; perform N-pt FFT (pass 3)
        mpfft     N,fftdat,ROM_TBL,ROM_LEN/2,4,1    ; perform N-pt FFT (pass 4)
                                            ; (w/ half-cycle tables)

;       move      #-1,m0            ; linear access
;       move      #1,n0             ; w/ offset inc of 1
        move      #0,m0             ; bit-reverse access
        move      #(N/2),n0         ; w/ offset inc associated w/ bit-rev access

        move      #fftdat,r0        ; initialize output data ptr register
        do        #N,lp3            ; transfer each of the output results
        put_OUT   x:(r0)+n0         ; to next PE in the array.  First loop
lp3                                 ; for all of the real parts of the results,
```

```
        move      #fftdat,r0      ; initialize output data ptr register
        do        #N,lp4          ; then loop for all of the imag parts
        put_OUT   y:(r0)+n0       ; (both with normal output ordering)
lp4
        jmp       LP              ; continue infinite loop
        stop
        end       START
```

```
        title     "(FFT32_1.ASM) FFT test for initial stage PEs (N=32)"

        include   'OP_SYS.INC'      ; incl operating system equates & macros

N       equ       32
        org       x:$0
fftdat  ds        N                 ; set aside buffer space for N pts (real)
        org       y:$0
        ds        N                 ; set aside buffer space for N pts (imag)

        org       p:$100
START
        init_PE                     ; init this PE

LP      move      #-1,m0            ; linear access
        move      #1,n0             ; w/ offset inc of 1
;       move      #0,m0             ; bit-reverse access
;       move      #(N/2),n0         ; w/ offset inc associated w/ bit-rev access

        move      #fftdat,r0        ; initialize input data ptr register
        do        #N,lp1            ; transfer each of the input values
        get_IN    d0.s              ; from previous input/PE.  First loop
        move      d0.s,x:(r0)+n0    ; for all of the real parts of the results,
lp1
        move      #fftdat,r0        ; initialize input data ptr register
        do        #N,lp2            ; then loop for all of the imag parts
        get_IN    d0.s              ; (both with normal output ordering)
        move      d0.s,y:(r0)+n0
lp2
                                              ; multipass FFT testing:
        mpfft     N,fftdat,ROM_TBL,ROM_LEN/2,1,2   ; perform N-pt FFT (pass 1-2)
;       mpfft     N,fftdat,ROM_TBL,ROM_LEN/2,3,1   ; perform N-pt FFT (pass 3)
;       mpfft     N,fftdat,ROM_TBL,ROM_LEN/2,4,2   ; perform N-pt FFT (pass 4-5)
                                              ; (w/ half-cycle tables)

        move      #-1,m0            ; linear access
        move      #1,n0             ; w/ offset inc of 1
;       move      #0,m0             ; bit-reverse access
;       move      #(N/2),n0         ; w/ offset inc associated w/ bit-rev access

        move      #fftdat,r0        ; initialize output data ptr register
        do        #N,lp3            ; transfer each of the output results
        nop                         ; to next PE in the array.  First loop
        put_PE    x:(r0)+n0         ; for all of the real parts of the results,
        nop                         ; (NOPs for DO instruction restrictions)
lp3
        move      #fftdat,r0        ; initialize output data ptr register
        do        #N,lp4            ; then loop for all of the imag parts
        nop                         ; (both with normal output ordering)
```

```
        put_PE    y:(r0)+n0
        nop                          ; (NOPs for DO instruction restrictions)
lp4
        jmp       LP                 ; continue infinite loop
        stop
        end       START
```

```
        title     "(FFT32_2.ASM) FFT test for intermed stage PEs (N=32)"

        include   'OP_SYS.INC'     ; incl operating system equates & macros

N       equ       32
        org       x:$0
fftdat  ds        N                ; set aside buffer space for N pts (real)
        org       y:$0
        ds        N                ; set aside buffer space for N pts (imag)

        org       p:$100
START
        init_PE                    ; init this PE

LP      move      #-1,m0           ; linear access
        move      #1,n0            ; w/ offset inc of 1
;       move      #0,m0            ; bit-reverse access
;       move      #(N/2),n0        ; w/ offset inc associated w/ bit-rev access

        move      #fftdat,r0       ; initialize input data ptr register
        do        #N,lp1           ; transfer each of the input values
        nop                        ; from previous input/PE.  First loop
        get_PE    d0.s             ; for all of the real parts of the results,
        move      d0.s,x:(r0)+n0
        nop                        ; (NOPs for DO instruction restrictions)
lp1
        move      #fftdat,r0       ; initialize input data ptr register
        do        #N,lp2           ; then loop for all of the imag parts
        nop                        ; (both with normal output ordering)
        get_PE    d0.s
        move      d0.s,y:(r0)+n0
        nop                        ; (NOPs for DO instruction restrictions)
lp2
                                                 ; multipass FFT testing:
;       mpfft     N,fftdat,ROM_TBL,ROM_LEN/2,1,2   ; perform N-pt FFT (pass 1-2)
        mpfft     N,fftdat,ROM_TBL,ROM_LEN/2,3,1   ; perform N-pt FFT (pass 3)
;       mpfft     N,fftdat,ROM_TBL,ROM_LEN/2,4,2   ; perform N-pt FFT (pass 4-5)
                                                 ; (w/ half-cycle tables)

        move      #-1,m0           ; linear access
        move      #1,n0            ; w/ offset inc of 1
;       move      #0,m0            ; bit-reverse access
;       move      #(N/2),n0        ; w/ offset inc associated w/ bit-rev access

        move      #fftdat,r0       ; initialize output data ptr register
        do        #N,lp3           ; transfer each of the output results
        nop                        ; to next PE in the array.  First loop
        put_PE    x:(r0)+n0        ; for all of the real parts of the results,
        nop                        ; (NOPs for DO instruction restrictions)
```

```
lp3
        move    #fftdat,r0      ; initialize output data ptr register
        do      #N,lp4          ; then loop for all of the imag parts
        nop                     ; (both with normal output ordering)
        put_PE  y:(r0)+n0
        nop                     ; (NOPs for DO instruction restrictions)
lp4
        jmp     LP              ; continue infinite loop
        stop
        end     START
```

```
        title     "(FFT32_3.ASM) FFT test for final stage PEs (N=32)"

        include   'OP_SYS.INC'      ; incl operating system equates & macros

N       equ       32
        org       x:$0
fftdat  ds        N                 ; set aside buffer space for N pts (real)
        org       y:$0
        ds        N                 ; set aside buffer space for N pts (imag)

        org       p:$100
START
        init_PE                     ; init this PE

LP      move      #-1,m0            ; linear access
        move      #1,n0             ; w/ offset inc of 1
;       move      #0,m0             ; bit-reverse access
;       move      #(N/2),n0         ; w/ offset inc associated w/ bit-rev access

        move      #fftdat,r0        ; initialize input data ptr register
        do        #N,lp1            ; transfer each of the input values
        nop                         ; from previous input/PE.  First loop
        get_PE    d0.s              ; for all of the real parts of the results,
        move      d0.s,x:(r0)+n0
        nop                         ; (NOPs for DO instruction restrictions
lp1
        move      #fftdat,r0        ; initialize input data ptr register
        do        #N,lp2            ; then loop for all of the imag parts
        nop                         ; (both with normal output ordering)
        get_PE    d0.s
        move      d0.s,y:(r0)+n0
        nop                         ; (NOPs for DO instruction restrictions)
lp2
                                                  ; multipass FFT testing:
;       mpfft     N,fftdat,ROM_TBL,ROM_LEN/2,1,2  ; perform N-pt FFT (pass 1-2)
;       mpfft     N,fftdat,ROM_TBL,ROM_LEN/2,3,1  ; perform N-pt FFT (pass 3)
        mpfft     N,fftdat,ROM_TBL,ROM_LEN/2,4,2  ; perform N-pt FFT (pass 4-5)
                                                  ; (w/ half-cycle tables)

;       move      #-1,m0            ; linear access
;       move      #1,n0             ; w/ offset inc of 1
        move      #0,m0             ; bit-reverse access
        move      #(N/2),n0         ; w/ offset inc associated w/ bit-rev access

        move      #fftdat,r0        ; initialize output data ptr register
        do        #N,lp3            ; transfer each of the output results
        put_OUT   x:(r0)+n0         ; to next PE in the array.  First loop
lp3                                 ; for all of the real parts of the results,
```

```
        move      #fftdat,r0      ; initialize output data ptr register
        do        #N,lp4          ; then loop for all of the imag parts
        put_OUT   y:(r0)+n0       ; (both with normal output ordering)
lp4
        jmp       LP              ; continue infinite loop
        stop
        end       START
```

```
        title     "(FFT64_1.ASM) FFT test for initial stage PEs (N=64)"

        include   'OP_SYS.INC'      ; incl operating system equates & macros

N       equ       64
        org       x:$0
fftdat  ds        N                 ; set aside buffer space for N pts (real)
        org       y:$0
        ds        N                 ; set aside buffer space for N pts (imag)

        org       p:$100
START
        init_PE                     ; init this PE

LP      move      #-1,m0            ; linear access
        move      #1,n0             ; w/ offset inc of 1
;       move      #0,m0             ; bit-reverse access
;       move      #(N/2),n0         ; w/ offset inc associated w/ bit-rev access

        move      #fftdat,r0        ; initialize input data ptr register
        do        #N,lp1            ; transfer each of the input values
        get_IN    d0.s              ; from previous input/PE.  First loop
        move      d0.s,x:(r0)+n0    ; for all of the real parts of the results,
lp1
        move      #fftdat,r0        ; initialize input data ptr register
        do        #N,lp2            ; then loop for all of the imag parts
        get_IN    d0.s              ; (both with normal output ordering)
        move      d0.s,y:(r0)+n0
lp2
                                                  ; multipass FFT testing:
        mpfft     N,fftdat,ROM_TBL,ROM_LEN/2,1,2  ; perform N-pt FFT (pass 1-2)
;       mpfft     N,fftdat,ROM_TBL,ROM_LEN/2,3,2  ; perform N-pt FFT (pass 3-4)
;       mpfft     N,fftdat,ROM_TBL,ROM_LEN/2,5,2  ; perform N-pt FFT (pass 5-6)
                                                  ; (w/ half-cycle tables)

        move      #-1,m0            ; linear access
        move      #1,n0             ; w/ offset inc of 1
;       move      #0,m0             ; bit-reverse access
;       move      #(N/2),n0         ; w/ offset inc associated w/ bit-rev access

        move      #fftdat,r0        ; initialize output data ptr register
        do        #N,lp3            ; transfer each of the output results
        nop                         ; to next PE in the array.  First loop
        put_PE    x:(r0)+n0         ; for all of the real parts of the results,
        nop                         ; (NOPs for DO instruction restrictions)
lp3
        move      #fftdat,r0        ; initialize output data ptr register
        do        #N,lp4            ; then loop for all of the imag parts
        nop                         ; (both with normal output ordering)
```

```
        put_PE   y:(r0)+n0
        nop                        ; (NOPs for DO instruction restrictions)
lp4
        jmp      LP                ; continue infinite loop
        stop
        end      START
```

```
        title     "(FFT64_2.ASM) FFT test for intermed stage PEs (N=64)"

        include   'OP_SYS.INC'      ; incl operating system equates & macros

N       equ       64
        org       x:$0
fftdat  ds        N                 ; set aside buffer space for N pts (real)
        org       y:$0
        ds        N                 ; set aside buffer space for N pts (imag)

        org       p:$100
START
        init_PE                     ; init this PE

LP      move      #-1,m0            ; linear access
        move      #1,n0             ; w/ offset inc of 1
;       move      #0,m0             ; bit-reverse access
;       move      #(N/2),n0         ; w/ offset inc associated w/ bit-rev access

        move      #fftdat,r0        ; initialize input data ptr register
        do        #N,lp1            ; transfer each of the input values
        nop                         ; from previous input/PE.  First loop
        get_PE    d0.s              ; for all of the real parts of the results,
        move      d0.s,x:(r0)+n0
        nop                         ; (NOPs for DO instruction restrictions)
lp1
        move      #fftdat,r0        ; initialize input data ptr register
        do        #N,lp2            ; then loop for all of the imag parts
        nop                         ; (both with normal output ordering)
        get_PE    d0.s
        move      d0.s,y:(r0)+n0
        nop                         ; (NOPs for DO instruction restrictions)
lp2
                                              ; multipass FFT testing:
;       mpfft     N,fftdat,ROM_TBL,ROM_LEN/2,1,2    ; perform N-pt FFT (pass 1-2)
        mpfft     N,fftdat,ROM_TBL,ROM_LEN/2,3,2    ; perform N-pt FFT (pass 3-4)
;       mpfft     N,fftdat,ROM_TBL,ROM_LEN/2,5,2    ; perform N-pt FFT (pass 5-6)
                                              ; (w/ half-cycle tables)

        move      #-1,m0            ; linear access
        move      #1,n0             ; w/ offset inc of 1
;       move      #0,m0             ; bit-reverse access
;       move      #(N/2),n0         ; w/ offset inc associated w/ bit-rev access

        move      #fftdat,r0        ; initialize output data ptr register
        do        #N,lp3            ; transfer each of the output results
        nop                         ; to next PE in the array.  First loop
        put_PE    x:(r0)+n0         ; for all of the real parts of the results,
        nop                         ; (NOPs for DO instruction restrictions)
```

```
lp3
        move     #fftdat,r0        ; initialize output data ptr register
        do       #N,lp4            ; then loop for all of the imag parts
        nop                        ; (both with normal output ordering)
        put_PE   y:(r0)+n0
        nop                        ; (NOPs for DO instruction restrictions)
lp4
        jmp      LP                ; continue infinite loop
        stop
        end      START
```

```
        title     "(FFT64_3.ASM) FFT test for final stage PEs (N=64)"

        include   'OP_SYS.INC'     ; incl operating system equates & macros

N       equ       64
        org       x:$0
fftdat  ds        N                ; set aside buffer space for N pts (real)
        org       y:$0
        ds        N                ; set aside buffer space for N pts (imag)

        org       p:$100
START
        init_PE                    ; init this PE

LP      move      #-1,m0           ; linear access
        move      #1,n0            ; w/ offset inc of 1
;       move      #0,m0            ; bit-reverse access
;       move      #(N/2),n0        ; w/ offset inc associated w/ bit-rev access

        move      #fftdat,r0       ; initialize input data ptr register
        do        #N,lp1           ; transfer each of the input values
        nop                        ; from previous input/PE. First loop
        get_PE    d0.s             ; for all of the real parts of the results,
        move      d0.s,x:(r0)+n0
        nop                        ; (NOPs for DO instruction restrictions
lp1
        move      #fftdat,r0       ; initialize input data ptr register
        do        #N,lp2           ; then loop for all of the imag parts
        nop                        ; (both with normal output ordering)
        get_PE    d0.s
        move      d0.s,y:(r0)+n0
        nop                        ; (NOPs for DO instruction restrictions)
lp2
                                                   ; multipass FFT testing:
;       mpfft     N,fftdat,ROM_TBL,ROM_LEN/2,1,2   ; perform N-pt FFT (pass 1-2)
;       mpfft     N,fftdat,ROM_TBL,ROM_LEN/2,3,2   ; perform N-pt FFT (pass 3-4)
        mpfft     N,fftdat,ROM_TBL,ROM_LEN/2,5,2   ; perform N-pt FFT (pass 5-6)
                                                   ; (w/ half-cycle tables)

;       move      #-1,m0           ; linear access
;       move      #1,n0            ; w/ offset inc of 1
        move      #0,m0            ; bit-reverse access
        move      #(N/2),n0        ; w/ offset inc associated w/ bit-rev access

        move      #fftdat,r0       ; initialize output data ptr register
        do        #N,lp3           ; transfer each of the output results
        put_OUT   x:(r0)+n0        ; to next PE in the array. First loop
lp3                                ; for all of the real parts of the results,
```

```
        move     #fftdat,r0        ; initialize output data ptr register
        do       #N,lp4            ; then loop for all of the imag parts
        put_OUT  y:(r0)+n0         ; (both with normal output ordering)
lp4
        jmp      LP                ; continue infinite loop
        stop
        end      START
```

```
        title     "(FFT128_1.ASM) FFT test for initial stage PEs (N=128)"

        include   'OP_SYS.INC'      ; incl operating system equates & macros

N       equ       128
        org       x:$0
fftdat  ds        N                 ; set aside buffer space for N pts (real)
        org       y:$0
        ds        N                 ; set aside buffer space for N pts (imag)

        org       p:$100
START
        init_PE                     ; init this PE

LP      move      #-1,m0            ; linear access
        move      #1,n0             ; w/ offset inc of 1
;       move      #0,m0             ; bit-reverse access
;       move      #(N/2),n0         ; w/ offset inc associated w/ bit-rev access

        move      #fftdat,r0        ; initialize input data ptr register
        do        #N,lp1            ; transfer each of the input values
        get_IN    d0.s              ; from previous input/PE.  First loop
        move      d0.s,x:(r0)+n0    ; for all of the real parts of the results,
lp1
        move      #fftdat,r0        ; initialize input data ptr register
        do        #N,lp2            ; then loop for all of the imag parts
        get_IN    d0.s              ; (both with normal output ordering)
        move      d0.s,y:(r0)+n0
lp2
                                                      ; multipass FFT testing:
        mpfft     N,fftdat,ROM_TBL,ROM_LEN/2,1,3   ; perform N-pt FFT (pass 1-3)
;       mpfft     N,fftdat,ROM_TBL,ROM_LEN/2,4,2   ; perform N-pt FFT (pass 4-5)
;       mpfft     N,fftdat,ROM_TBL,ROM_LEN/2,6,2   ; perform N-pt FFT (pass 6-7)
                                                      ; (w/ half-cycle tables)

        move      #-1,m0            ; linear access
        move      #1,n0             ; w/ offset inc of 1
;       move      #0,m0             ; bit-reverse access
;       move      #(N/2),n0         ; w/ offset inc associated w/ bit-rev access

        move      #fftdat,r0        ; initialize output data ptr register
        do        #N,lp3            ; transfer each of the output results
        nop                         ; to next PE in the array.  First loop
        put_PE    x:(r0)+n0         ; for all of the real parts of the results,
        nop                         ; (NOPs for DO instruction restrictions)
lp3
        move      #fftdat,r0        ; initialize output data ptr register
        do        #N,lp4            ; then loop for all of the imag parts
        nop                         ; (both with normal output ordering)
```

```
        put_PE   y:(r0)+n0
        nop                          ; (NOPs for DO instruction restrictions)
lp4
        jmp      LP                  ; continue infinite loop
        stop
        end      START
```

```
        title       "(FFT128_2.ASM) FFT test for intermed stage PEs (N=128)"

        include     'OP_SYS.INC'      ; incl operating system equates & macros

N       equ         128
        org         x:$0
fftdat  ds          N                 ; set aside buffer space for N pts (real)
        org         y:$0
        ds          N                 ; set aside buffer space for N pts (imag)

        org         p:$100
START
        init_PE                       ; init this PE

LP      move        #-1,m0            ; linear access
        move        #1,n0             ; w/ offset inc of 1
;       move        #0,m0             ; bit-reverse access
;       move        #(N/2),n0         ; w/ offset inc associated w/ bit-rev access

        move        #fftdat,r0        ; initialize input data ptr register
        do          #N,lp1            ; transfer each of the input values
        nop                           ; from previous input/PE.  First loop
        get_PE      d0.s              ; for all of the real parts of the results,
        move        d0.s,x:(r0)+n0
        nop                           ; (NOPs for DO instruction restrictions)
lp1
        move        #fftdat,r0        ; initialize input data ptr register
        do          #N,lp2            ; then loop for all of the imag parts
        nop                           ; (both with normal output ordering)
        get_PE      d0.s
        move        d0.s,y:(r0)+n0
        nop                           ; (NOPs for DO instruction restrictions)
lp2
                                                          ; multipass FFT testing:
;       mpfft       N,fftdat,ROM_TBL,ROM_LEN/2,1,3   ; perform N-pt FFT (pass 1-3)
        mpfft       N,fftdat,ROM_TBL,ROM_LEN/2,4,2   ; perform N-pt FFT (pass 4-5)
;       mpfft       N,fftdat,ROM_TBL,ROM_LEN/2,6,2   ; perform N-pt FFT (pass 6-7)
                                                          ; (w/ half-cycle tables)

        move        #-1,m0            ; linear access
        move        #1,n0             ; w/ offset inc of 1
;       move        #0,m0             ; bit-reverse access
;       move        #(N/2),n0         ; w/ offset inc associated w/ bit-rev access

        move        #fftdat,r0        ; initialize output data ptr register
        do          #N,lp3            ; transfer each of the output results
        nop                           ; to next PE in the array.  First loop
        put_PE      x:(r0)+n0         ; for all of the real parts of the results,
        nop                           ; (NOPs for DO instruction restrictions)
```

```
lp3
        move     #fftdat,r0      ; initialize output data ptr register
        do       #N,lp4          ; then loop for all of the imag parts
        nop                      ; (both with normal output ordering)
        put_PE   y:(r0)+n0
        nop                      ; (NOPs for DO instruction restrictions)
lp4
        jmp      LP              ; continue infinite loop
        stop
        end      START
```

```
        title       "(FFT128_3.ASM) FFT test for final stage PEs (N=128)"

        include     'OP_SYS.INC'      ; incl operating system equates & macros

N       equ         128
        org         x:$0
fftdat  ds          N                 ; set aside buffer space for N pts (real)
        org         y:$0
        ds          N                 ; set aside buffer space for N pts (imag)

        org         p:$100
START
        init_PE                       ; init this PE

LP      move        #-1,m0            ; linear access
        move        #1,n0             ; w/ offset inc of 1
;       move        #0,m0             ; bit-reverse access
;       move        #(N/2),n0         ; w/ offset inc associated w/ bit-rev access

        move        #fftdat,r0        ; initialize input data ptr register
        do          #N,lp1            ; transfer each of the input values
        nop                           ; from previous input/PE.  First loop
        get_PE      d0.s              ; for all of the real parts of the results,
        move        d0.s,x:(r0)+n0
        nop                           ; (NOPs for DO instruction restrictions
lp1
        move        #fftdat,r0        ; initialize input data ptr register
        do          #N,lp2            ; then loop for all of the imag parts
        nop                           ; (both with normal output ordering)
        get_PE      d0.s
        move        d0.s,y:(r0)+n0
        nop                           ; (NOPs for DO instruction restrictions)
lp2
                                                  ; multipass FFT testing:
;       mpfft       N,fftdat,ROM_TBL,ROM_LEN/2,1,3   ; perform N-pt FFT (pass 1-3)
;       mpfft       N,fftdat,ROM_TBL,ROM_LEN/2,4,2   ; perform N-pt FFT (pass 4-5)
        mpfft       N,fftdat,ROM_TBL,ROM_LEN/2,6,2   ; perform N-pt FFT (pass 6-7)
                                                  ; (w/ half-cycle tables)

;       move        #-1,m0            ; linear access
;       move        #1,n0             ; w/ offset inc of 1
        move        #0,m0             ; bit-reverse access
        move        #(N/2),n0         ; w/ offset inc associated w/ bit-rev access

        move        #fftdat,r0        ; initialize output data ptr register
        do          #N,lp3            ; transfer each of the output results
        put_OUT     x:(r0)+n0         ; to next PE in the array.  First loop
lp3                                   ; for all of the real parts of the results,
```

```
        move     #fftdat,r0      ; initialize output data ptr register
        do       #N,lp4          ; then loop for all of the imag parts
        put_OUT  y:(r0)+n0       ; (both with normal output ordering)
lp4
        jmp      LP              ; continue infinite loop
        stop
        end      START
```

```
        title    "(FFT256_1.ASM) FFT test for initial stage PEs (N=256)"

        include  'OP_SYS.INC'      ; incl operating system equates & macros

N       equ      256
        org      x:$0
fftdat  ds       N                 ; set aside buffer space for N pts (real)
        org      y:$0
        ds       N                 ; set aside buffer space for N pts (imag)

        org      p:$100
START
        init_PE                    ; init this PE

LP      move     #-1,m0            ; linear access
        move     #1,n0             ; w/ offset inc of 1
;       move     #0,m0             ; bit-reverse access
;       move     #(N/2),n0         ; w/ offset inc associated w/ bit-rev access

        move     #fftdat,r0        ; initialize input data ptr register
        do       #N,lp1            ; transfer each of the input values
        get_IN   d0.s              ; from previous input/PE.  First loop
        move     d0.s,x:(r0)+n0    ; for all of the real parts of the results,
lp1
        move     #fftdat,r0        ; initialize input data ptr register
        do       #N,lp2            ; then loop for all of the imag parts
        get_IN   d0.s              ; (both with normal output ordering)
        move     d0.s,y:(r0)+n0
lp2
                                             ; multipass FFT testing:
        mpfft    N,fftdat,ROM_TBL,ROM_LEN/2,1,3    ; perform N-pt FFT (pass 1-3)
;       mpfft    N,fftdat,ROM_TBL,ROM_LEN/2,4,2    ; perform N-pt FFT (pass 4-5)
;       mpfft    N,fftdat,ROM_TBL,ROM_LEN/2,6,3    ; perform N-pt FFT (pass 6-8)
                                             ; (w/ half-cycle tables)

        move     #-1,m0            ; linear access
        move     #1,n0             ; w/ offset inc of 1
;       move     #0,m0             ; bit-reverse access
;       move     #(N/2),n0         ; w/ offset inc associated w/ bit-rev access

        move     #fftdat,r0        ; initialize output data ptr register
        do       #N,lp3            ; transfer each of the output results
        nop                        ; to next PE in the array.  First loop
        put_PE   x:(r0)+n0         ; for all of the real parts of the results,
        nop                        ; (NOPs for DO instruction restrictions)
lp3
        move     #fftdat,r0        ; initialize output data ptr register
        do       #N,lp4            ; then loop for all of the imag parts
        nop                        ; (both with normal output ordering)
```

```
            put_PE    y:(r0)+n0
            nop                        ; (NOPs for DO instruction restrictions)
lp4
            jmp       LP               ; continue infinite loop
            stop
            end       START
```

```
        title     "(FFT256_2.ASM) FFT test for intermed stage PEs (N=256)"

        include   'OP_SYS.INC'      ; incl operating system equates & macros

N       equ       256
        org       x:$0
fftdat  ds        N                 ; set aside buffer space for N pts (real)
        org       y:$0
        ds        N                 ; set aside buffer space for N pts (imag)

        org       p:$100
START
        init_PE                     ; init this PE

LP      move      #-1,m0            ; linear access
        move      #1,n0             ; w/ offset inc of 1
;       move      #0,m0             ; bit-reverse access
;       move      #(N/2),n0         ; w/ offset inc associated w/ bit-rev access

        move      #fftdat,r0        ; initialize input data ptr register
        do        #N,lp1            ; transfer each of the input values
        nop                         ; from previous input/PE.  First loop
        get_PE    d0.s              ; for all of the real parts of the results,
        move      d0.s,x:(r0)+n0
        nop                         ; (NOPs for DO instruction restrictions)
lp1
        move      #fftdat,r0        ; initialize input data ptr register
        do        #N,lp2            ; then loop for all of the imag parts
        nop                         ; (both with normal output ordering)
        get_PE    d0.s
        move      d0.s,y:(r0)+n0
        nop                         ; (NOPs for DO instruction restrictions)
lp2
                                                        ; multipass FFT testing:
;       mpfft     N,fftdat,ROM_TBL,ROM_LEN/2,1,3   ; perform N-pt FFT (pass 1-3)
        mpfft     N,fftdat,ROM_TBL,ROM_LEN/2,4,2   ; perform N-pt FFT (pass 4-5)
;       mpfft     N,fftdat,ROM_TBL,ROM_LEN/2,6,3   ; perform N-pt FFT (pass 6-8)
                                                        ; (w/ half-cycle tables)

        move      #-1,m0            ; linear access
        move      #1,n0             ; w/ offset inc of 1
;       move      #0,m0             ; bit-reverse access
;       move      #(N/2),n0         ; w/ offset inc associated w/ bit-rev access

        move      #fftdat,r0        ; initialize output data ptr register
        do        #N,lp3            ; transfer each of the output results
        nop                         ; to next PE in the array.  First loop
        put_PE    x:(r0)+n0         ; for all of the real parts of the results,
        nop                         ; (NOPs for DO instruction restrictions)
```

```
lp3
        move      #fftdat,r0      ; initialize output data ptr register
        do        #N,lp4          ; then loop for all of the imag parts
        nop                       ; (both with normal output ordering)
        put_PE    y:(r0)+n0
        nop                       ; (NOPs for DO instruction restrictions)
lp4
        jmp       LP              ; continue infinite loop
        stop
        end       START
```

```
        title     "(FFT256_3.ASM) FFT test for final stage PEs (N=256)"

        include   'OP_SYS.INC'       ; incl operating system equates & macros

N       equ       256
        org       x:$0
fftdat  ds        N                  ; set aside buffer space for N pts (real)
        org       y:$0
        ds        N                  ; set aside buffer space for N pts (imag)

        org       p:$100
START
        init_PE                      ; init this PE

LP      move      #-1,m0             ; linear access
        move      #1,n0              ; w/ offset inc of 1
;       move      #0,m0              ; bit-reverse access
;       move      #(N/2),n0          ; w/ offset inc associated w/ bit-rev access

        move      #fftdat,r0         ; initialize input data ptr register
        do        #N,lp1             ; transfer each of the input values
        nop                          ; from previous input/PE.  First loop
        get_PE    d0.s               ; for all of the real parts of the results,
        move      d0.s,x:(r0)+n0
        nop                          ; (NOPs for DO instruction restrictions
lp1
        move      #fftdat,r0         ; initialize input data ptr register
        do        #N,lp2             ; then loop for all of the imag parts
        nop                          ; (both with normal output ordering)
        get_PE    d0.s
        move      d0.s,y:(r0)+n0
        nop                          ; (NOPs for DO instruction restrictions)
lp2
                                                  ; multipass FFT testing:
;       mpfft     N,fftdat,ROM_TBL,ROM_LEN/2,1,3  ; perform N-pt FFT (pass 1-3)
;       mpfft     N,fftdat,ROM_TBL,ROM_LEN/2,4,2  ; perform N-pt FFT (pass 4-5)
        mpfft     N,fftdat,ROM_TBL,ROM_LEN/2,6,3  ; perform N-pt FFT (pass 6-8)
                                                  ; (w/ half-cycle tables)

;       move      #-1,m0             ; linear access
;       move      #1,n0              ; w/ offset inc of 1
        move      #0,m0              ; bit-reverse access
        move      #(N/2),n0          ; w/ offset inc associated w/ bit-rev access

        move      #fftdat,r0         ; initialize output data ptr register
        do        #N,lp3             ; transfer each of the output results
        put_OUT   x:(r0)+n0          ; to next PE in the array.  First loop
lp3                                  ; for all of the real parts of the results,
```

```
        move     #fftdat,r0       ; initialize output data ptr register
        do       #N,lp4           ; then loop for all of the imag parts
        put_OUT  y:(r0)+n0        ; (both with normal output ordering)
lp4
        jmp      LP               ; continue infinite loop
        stop
        end      START
```

```
        title     "(FFT512_1.ASM) FFT test for initial stage PEs (N=512)"

        include   'OP_SYS.INC'      ; incl operating system equates & macros

N       equ       512
        org       x:$0
fftdat  ds        N                 ; set aside buffer space for N pts (real)
        org       y:$0
        ds        N                 ; set aside buffer space for N pts (imag)

        org       p:$100
START
        init_PE                     ; init this PE

LP      move      #-1,m0            ; linear access
        move      #1,n0             ; w/ offset inc of 1
;       move      #0,m0             ; bit-reverse access
;       move      #(N/2),n0         ; w/ offset inc associated w/ bit-rev access

        move      #fftdat,r0        ; initialize input data ptr register
        do        #N,lp1            ; transfer each of the input values
        get_IN    d0.s              ; from previous input/PE.  First loop
        move      d0.s,x:(r0)+n0    ; for all of the real parts of the results,
lp1
        move      #fftdat,r0        ; initialize input data ptr register
        do        #N,lp2            ; then loop for all of the imag parts
        get_IN    d0.s              ; (both with normal output ordering)
        move      d0.s,y:(r0)+n0
lp2
                                                        ; multipass FFT testing:
        mpfft     N,fftdat,ROM_TBL,ROM_LEN/2,1,3   ; perform N-pt FFT (pass 1-3)
;       mpfft     N,fftdat,ROM_TBL,ROM_LEN/2,4,3   ; perform N-pt FFT (pass 4-6)
;       mpfft     N,fftdat,ROM_TBL,ROM_LEN/2,7,3   ; perform N-pt FFT (pass 7-9)
                                                        ; (w/ half-cycle tables)

        move      #-1,m0            ; linear access
        move      #1,n0             ; w/ offset inc of 1
;       move      #0,m0             ; bit-reverse access
;       move      #(N/2),n0         ; w/ offset inc associated w/ bit-rev access

        move      #fftdat,r0        ; initialize output data ptr register
        do        #N,lp3            ; transfer each of the output results
        nop                         ; to next PE in the array.  First loop
        put_PE    x:(r0)+n0         ; for all of the real parts of the results,
        nop                         ; (NOPs for DO instruction restrictions)
lp3
        move      #fftdat,r0        ; initialize output data ptr register
        do        #N,lp4            ; then loop for all of the imag parts
        nop                         ; (both with normal output ordering)
```

```
          put_PE    y:(r0)+n0
          nop                        ; (NOPs for DO instruction restrictions)
lp4
          jmp       LP               ; continue infinite loop
          stop
          end       START
```

```
        title     "(FFT512_2.ASM) FFT test for intermed stage PEs (N=512)"

        include   'OP_SYS.INC'      ; incl operating system equates & macros

N       equ       512
        org       x:$0
fftdat  ds        N                 ; set aside buffer space for N pts (real)
        org       y:$0
        ds        N                 ; set aside buffer space for N pts (imag)

        org       p:$100
START
        init_PE                     ; init this PE

LP      move      #-1,m0            ; linear access
        move      #1,n0             ; w/ offset inc of 1
;       move      #0,m0             ; bit-reverse access
;       move      #(N/2),n0         ; w/ offset inc associated w/ bit-rev access

        move      #fftdat,r0        ; initialize input data ptr register
        do        #N,lp1            ; transfer each of the input values
        nop                         ; from previous input/PE.  First loop
        get_PE    d0.s              ; for all of the real parts of the results,
        move      d0.s,x:(r0)+n0
        nop                         ; (NOPs for DO instruction restrictions)
lp1
        move      #fftdat,r0        ; initialize input data ptr register
        do        #N,lp2            ; then loop for all of the imag parts
        nop                         ; (both with normal output ordering)
        get_PE    d0.s
        move      d0.s,y:(r0)+n0
        nop                         ; (NOPs for DO instruction restrictions)
lp2
                                              ; multipass FFT testing:
;       mpfft     N,fftdat,ROM_TBL,ROM_LEN/2,1,3   ; perform N-pt FFT (pass 1-3)
        mpfft     N,fftdat,ROM_TBL,ROM_LEN/2,4,3   ; perform N-pt FFT (pass 4-6)
;       mpfft     N,fftdat,ROM_TBL,ROM_LEN/2,7,3   ; perform N-pt FFT (pass 7-9)
                                              ; (w/ half-cycle tables)

        move      #-1,m0            ; linear access
        move      #1,n0             ; w/ offset inc of 1
;       move      #0,m0             ; bit-reverse access
;       move      #(N/2),n0         ; w/ offset inc associated w/ bit-rev access

        move      #fftdat,r0        ; initialize output data ptr register
        do        #N,lp3            ; transfer each of the output results
        nop                         ; to next PE in the array.  First loop
        put_PE    x:(r0)+n0         ; for all of the real parts of the results,
        nop                         ; (NOPs for DO instruction restrictions)
```

```
lp3
        move      #fftdat,r0       ; initialize output data ptr register
        do        #N,lp4           ; then loop for all of the imag parts
        nop                        ; (both with normal output ordering)
        put_PE    y:(r0)+n0
        nop                        ; (NOPs for DO instruction restrictions)
lp4
        jmp       LP               ; continue infinite loop
        stop
        end       START
```

```
        title     "(FFT512_3.ASM) FFT test for final stage PEs (N=512)"

        include   'OP_SYS.INC'     ; incl operating system equates & macros

N       equ       512
        org       x:$0
fftdat  ds        N                ; set aside buffer space for N pts (real)
        org       y:$0
        ds        N                ; set aside buffer space for N pts (imag)

        org       p:$100
START
        init_PE                    ; init this PE

LP      move      #-1,m0           ; linear access
        move      #1,n0            ; w/ offset inc of 1
;       move      #0,m0            ; bit-reverse access
;       move      #(N/2),n0        ; w/ offset inc associated w/ bit-rev access

        move      #fftdat,r0       ; initialize input data ptr register
        do        #N,lp1           ; transfer each of the input values
        nop                        ; from previous input/PE.  First loop
        get_PE    d0.s             ; for all of the real parts of the results,
        move      d0.s,x:(r0)+n0
        nop                        ; (NOPs for DO instruction restrictions)
lp1
        move      #fftdat,r0       ; initialize input data ptr register
        do        #N,lp2           ; then loop for all of the imag parts
        nop                        ; (both with normal output ordering)
        get_PE    d0.s
        move      d0.s,y:(r0)+n0
        nop                        ; (NOPs for DO instruction restrictions)
lp2
                                                 ; multipass FFT testing:
;       mpfft     N,fftdat,ROM_TBL,ROM_LEN/2,1,3   ; perform N-pt FFT (pass 1-3)
;       mpfft     N,fftdat,ROM_TBL,ROM_LEN/2,4,3   ; perform N-pt FFT (pass 4-6)
        mpfft     N,fftdat,ROM_TBL,ROM_LEN/2,7,3   ; perform N-pt FFT (pass 7-9)
                                                 ; (w/ half-cycle tables)

;       move      #-1,m0           ; linear access
;       move      #1,n0            ; w/ offset inc of 1
        move      #0,m0            ; bit-reverse access
        move      #(N/2),n0        ; w/ offset inc associated w/ bit-rev access

        move      #fftdat,r0       ; initialize output data ptr register
        do        #N,lp3           ; transfer each of the output results
        put_OUT   x:(r0)+n0        ; to next PE in the array.  First loop
lp3                                ; for all of the real parts of the results,
```

```
        move      #fftdat,r0       ; initialize output data ptr register
        do        #N,lp4           ; then loop for all of the imag parts
        put_OUT   y:(r0)+n0        ; (both with normal output ordering)
lp4
        jmp       LP               ; continue infinite loop
        stop
        end       START
```

BIBLIOGRAPHY

[ABBO90] Abbott, R.J., "Resourceful Systems for Fault Tolerance, Reliability, and Safety," *ACM Computing Surveys*, Vol. 22, No. 1, March 1990.

[ABRA87] Abraham, J.A., P. Banerjee, C-Y Chen, W.K. Fuchs, S-Y Kuo, and A.L.N. Reddy, "Fault Tolerance Techniques for Systolic Arrays," *IEEE Computer*, Vol. 20, No. 7, pp. 65-74, July 1987.

[ALLE85] Allen, J., "Computer Architecture for Digital Signal Processing," *Proceedings of the IEEE*, pp. 852-873, May 1985.

[ANDE81] Anderson, T., and P.A. Lee, *Fault Tolerance Principles and Practices*, Prentice-Hall International, London, 1981.

[ARMS81] Armstrong, J.R. and F.G. Gray, "Fault Diagnosis in Boolean n-Cube Array of Microprocessors," *IEEE Transactions on Computers*, pp. 590-595, August 1981.

[AUGU86] August Systems Data Sheet, *The CS-3001 Control Computer System Description*, August Systems, 18277 S.W. Boones Ferry Road, Tigard, Oregon, 97224-7673, 1986.

[BANE90] Banerjee, P., J.T. Rahmeh, C. Stunkel, V.S. Nair, K. Roy, V. Balasubramanian, and J.A. Abraham, "Algorithm-Based Fault Tolerance on a Hypercube Multiprocessor," *IEEE Transactions on Computers*, Vol. 39, No. 9, pp. 1132-1145, September 1990.

[BOWE82] Bowen, B.A. and W.R. Brown, *VLSI Systems Design for Digital Signal Processing*, Volume I, Prentice-Hall, Englewood Cliffs, New Jersey, 1982.

[BOWE85] Bowen, B.A. and W.R. Brown, *VLSI Systems Design for Digital Signal Processing*, Volume II, Prentice-Hall, Englewood Cliffs, New Jersey, 1985.

[BROO83] Broomell, G. and J.R. Heath, "Classification Categories and Historical Development of Circuit Switching Topologies," *ACM Computer Surveys*, Vol. 15, No. 2, June 1983.

[BRIG74] Brigham, E.O., *The Fast Fourier Transform*, Prentice-Hall, Englewood Cliffs, New Jersey, 1974.

[CART85] Carter, W.C., "Hardware Fault Tolerance," from T. Anderson, T., editor, *Resilient Computing Systems*, Collins Professional and Technical Books, London, England, pp. 11-63, 1985.

[CHEN78] Chen, L. and A. Avizienis, "N-Version Programming: A Fault Tolerant Approach to Reliability of Software Operation," *Digest of the 8th Annual International Symposium on Fault Tolerant Computing*, pp. 3-9, 1978.

[CHEN85] Chen, Y., and T. Chen, "DFT: Distributed Fault Tolerance - Analysis and Design," *Digest of the 15th Annual International Symposium on Fault-Tolerant Computing*, Ann Arbor, Michigan, June 1985.

[CHER91] Cheriton, D.R., H.A. Goosen, and P.D. Boyle, "Paradigm: A Highly Scalable Shared-Memory Multicomputer Architecture," *IEEE Computer*, Vol. 24, No.2, pp. 33-46, February 1991.

[CHOI85] Choi, Y. and M. Malek, "A Fault-Tolerant FFT Processor," *Digest of the 15th Annual International Symposium on Fault-Tolerant Computing*, Ann Arbor, Michigan, pp. 266-271, June 19-21, 1985.

[CLAR82] Clark, E.M., and C.N. Nickolaous, "Distributed Reconfiguration Strategies for Fault Tolerant Microprocessor Systems," *IEEE Transactions on Computers*, Vol. C-31, No. 8, August 1982.

[CRIS91] Cristian, Flavin, "Understanding Fault-Tolerant Distributed Systems," *Communications of the ACM*, Vol. 34, No. 2, pp. 57-78, February 1991.

[DALY73] Daly, W.M., A.L. Hopkins, and J.F. McKenna, "A Fault-Tolerant Digital Clocking System," *Digest of the 3rd Annual International Symposium on Fault Tolerant Computing*, pp. 17-22, June 1973.

[DAVI78] Davies, D. and J.F. Wakerly, "Synchronization and Matching in Redundant Systems," *IEEE Transactions on Computers*, Vol. C-27, No. 6, pp. 531-539, June 1978.

[DECE89] DeCegama, A.L., *Parallel Processing Architectures and VLSI Hardware*, Prentice-Hall, Englewood Cliffs, NJ, 1989.

[DEIT90] Deitel, H.M., *Operating Systems*, 2nd Edition, Addison-Wesley, Reading, MA, 1990.

[DESP78] Despain, A. and D. Patterson, "X-Tree: A Tree Structured Multiprocessor Computer Architecture," *Proceedings of the 5th Annual Symposium on Computer Architecture*, pp. 144-151, April 1978.

[DHIL87] Dhillon, B.S., *Reliability in Computer System Design*, Ablex Publishing, Norwood, New Jersey, 1987.

[DIMM85] Dimmer, C.I., "The Tandem Non-Stop system," from T. Anderson, T., editor, *Resilient Computing Systems*, Collins Professional and Technical Books, London, England, pp. 178-196, 1985.

[DSP89] *DSP96002 IEEE Floating-Point Dual-Port Processor User's Manual*, Publication No. DSP96002UM/AD, Motorola Inc., Austin, Texas, 1989.

[FLYN66] Flynn, J.J., "Very High-Speed Computing Systems," *Proceedings of the IEEE*, pp. 1901-1909, December, 1966.

[FORT84] Fortes, J.A.B., and C.S. Raghavendra, "Dynamically Reconfigurable Fault-Tolerant Array Processors," *Digest of the 14th Annual International Conference on Fault-Tolerant Computing*, Kissimmee, FL, pp. 386-392, June 1984.

[GARC89] Garcia-Molina, H. and F. Pittelli, "Implementing Reliable Distributed Supercomputing Systems," from S.P. Kartashev and S.I. Kartashev, editors, *Designing and Programming Modern Computer Systems Volume III*, Prentice-Hall, Englewood Cliffs, New Jersey, pp. 191-250, 1989.

[GEOR88] George, A.D. and F.O. Simons, Jr., "The Evolution and Design of DSP VLSI Hardware," an invited paper presented at *The International Conference on Science and Technology*, Dubrounik, Yugoslavia, June 1988.

[GEOR89a] George, A.D., "Transputers Applied to Digital Filter Design," *Proceedings of the Southeastern Symposium on System Theory (SSST)*, Tallahassee, Florida, pp. 399-403, March 1989.

[GEOR89b] George, A.D. and F.O. Simons, Jr., "VLSI Hardware Alternatives in DSP," *Proceedings of the 12th Annual Pittsburgh Conference on Modeling and Simulation*, Pittsburgh, Pennsylvania, pp. 1559-1563, May 1989.

[GEOR91] George, A.D., "A Fault-Tolerant Computing System for Digital Signal Processing," Ph.D. Dissertation, Computer Science Department, Florida State University, August 1991.

[GEOR92] George, A.D., "DSP96002 versus i860: Digital Signal Processing Applications," *Proceedings of the Souteastern Symposium on System Theory (SSST)*, Greensboro, North Carolina, pp. 110-114, March 1992.

[GILB85] Gilbert, B.K., T.M. Kinter, D.J. Schwab, B.A. Naused, L.M. Krueger, W.Van Nurden, and R. Zucca, "Signal Processing in High-Data-Rate Environments: Design Trade-Offs in the Exploitation of Parallel Architectures and Fast System Clock Rates," from S.Y. Kung, H.J. Whitehouse, and T. Kailath, editors, *VLSI and Modern Signal Processing*, Prentice-Hall, Englewood Cliffs, New Jersey, pp. 451-473, 1985.

[GUPT86] Gupta, R., A. Zorat, and I.V. Ramakrishnan, "A Fault-Tolerant Multipipeline Architecture," *Digest of the 16th Annual International Symposium on Fault-Tolerant Computing*, Vienna, Austria, pp. 350-355, July 1-4, 1986.

[HAWK85] Hawkes, L.W., "A Regular Fault-tolerant Architecture for Interconnection Networks", *IEEE Transactions on Computers*, C-34(7), pp. 677-680, July 1985.

[HAYE76] Hayes, J.P., "A Graph Model for Fault-Tolerant Computing Systems," *IEEE Transactions on Computers*, Vol. 25, No. 9, pp. 875-884, September 1976.

[HWAN84] Hwang, K. and F.A. Briggs, *Computer Architecture and Parallel Processing*, McGraw-Hill, New York, NY, 1984.

[IEEE85] ANSI/IEEE Standard 754-1985: IEEE Standard for Binary Floating-Point Arithmetic, IEEE Service Center, Piscataway, New Jersey, 1985.

[JOHN84] Johnson, B.W., "Fault-Tolerant Microprocessor-based Systems," *IEEE Micro*, Vol. 4, No. 6, pp. 6-21, December 1984.

[JOHN89a] Johnson, B.W., *Design and Analysis of Fault-Tolerant Digital Systems*, Addison-Wesley Publishing Company, Reading, Massachusetts, 1989.

[JOHN89b] Johnson, J.R., *Introduction to Digital Signal Processing*, Prentice-Hall, Englewood Cliffs, New Jersey, 1989.

[JOU85] Jou, J. and J.A. Abraham, "Fault-Tolerant FFT Networks," *Digest of the 15th Annual International Symposium on Fault-Tolerant Computing*, Ann Arbor, Michigan, pp. 338-343, June 19-21, 1985.

[KAWA80] Kawakubo, K., H. Nakamura, and I. Okumura, "The Architecture of a Fail-Safe and Fault-Tolerant Computer for a Railway Signaling Device," *Digest of the 10th Annual International Symposium on Fault-Tolerant Computing*, Kyoto, Japan, pp. 372-374, October 1-3, 1980.

[KLOK89a] Kloker, K., B. Lindsley, S. Liberman, P. Marino, E. Rushinek, and G.D. Hillman, "The Motorola DSP96002 IEEE Floating-Point Digital Signal Processor," *Proceedings of the IEEE International Conference on ASSP*, May 1989.

[KLOK89b] Kloker, K., B. Lindsley, N. Baron, and G.R.L. Sohie, "Efficient FFT Implementation on an IEEE Floating-Point Digital Signal Processor," *Proceedings of the IEEE International Conference on ASSP*, May 1989.

[KOPE85] Kopetz, H., "Resilient Real-Time Systems," from T. Anderson, editor, *Resilient Computing Systems*, Collins Professional and Technical Books, London, England, pp. 91-101, 1985.

[KRIS85] Krishna, C.M., K.G. Shin, and R.W. Butler, "Ensuring Fault Tolerance of Phase-Locked Clocks," *IEEE Transactions on Computers*, Vol. C-34, No. 8, pp. 752-756, August 1985.

[KUHL80] Kuhl, J., "Fault Diagnosis in Computing Networks," Ph.D. dissertation, Department of Electrical and Computer Engineering, University of Iowa, August 1980.

[KUNG82a] Kung, H.T., "Why Systolic Architectures?," *IEEE Computer*, Vol. 15, No. 1, pp. 37-46, January 1982.

[KUNG82b] Kung, S.Y., K.S. Arun, R.J. Gal-Ezer, and D.V. Bhaskar Rao, "Wavefront Array Processor: Language, Architecture, and Applications," *IEEE Transactions on Computers, Special Issue on*

Parallel and Distributed Computers, Vol. C-31, No. 11, pp. 1054-1066, November 1982.

[KWAN81] Kwan, C.L., and S. Toida, "Optimal Fault-Tolerant Realizations of Some Classes of Hierarchical Tree Systems," *Digest of the 11th Annual International Symposium on Fault-Tolerant Computing*, Portland, Maine, pp. 176-178, June 1981.

[LAMP81] Lampson, B.W., *Distributed Systems - Architecture and Implementation*, Springer-Verlag, Berlin, Germany, 1981.

[LANG88] Lang, R.L., M. Dharssi, F.M. Longstaff, P.S. Longstaff, P.A.S. Metford, and M.T. Rimmer, "An Optimum Parallel Architecture for High-Speed Real-Time Digital Signal Processing," *IEEE Computer*, Vol. 21, No. 2, pp. 47-57, February 1988.

[LAPR85] Laprie, J.C., "Dependable Computing and Fault Tolerance: Concepts and Terminology," *Proceedings of the 15th Annual International Symposium on Fault-Tolerant Computing*, Ann Arbor, Michigan, pp. 2-11, June 1985.

[LAWR87] Lawrence, P.D. and K. Mauch, *Real-Time Microcomputer System Design*, McGraw-Hill, New York, NY, 1987.

[LEE85] Lee, P.A. and T. Anderson, "Design Fault Tolerance," from T. Anderson, editor, *Resilient Computing Systems*, Collins Professional and Technical Books, London, England, pp. 64-77, 1985.

[LEE88] Lee, Edward A., "Programmable DSP Architectures: Part I," *IEEE ASSP Magazine*, pp. 4-19, October 1988.

[LEE89] Lee, Edward A., "Programmable DSP Architectures: Part II," *IEEE ASSP Magazine*, pp. 4-14, January 1989.

[LEE90] Lee, P.A. and T. Anderson, *Fault Tolerance Principles and Practice*, Springer-Verlag/Wien, New York, NY, 1990.

[MAEK87] Maekawa, M., A.E. Oldehoeft, and R.R. Oldehoeft, *Operating Systems: Advanced Concepts*, Benjamin Cummings, Menlo Park, CA, 1987.

[McCO81] McConnel, S.R. and D.P. Siewiorek, "Synchronization and Voting," *IEEE Transactions on Computers*, Vol. C-30, No. 2, pp. 161-164, February 1981.

[MEYE71] Meyer, J.F., "Fault Tolerant Sequential Machines," *IEEE Transactions on Computers*, Vol. C-20, No. 10, pp. 1167-1177, October 1971.

[MOTO89] *Motorola DSP96002 Digital Signal Processor Simulator Reference Manual*, Motorola Inc., Austin, Texas, 1989.

[MOTT86] Mott, J.L., A. Kandel, and T.P. Baker, *Discrete Mathematics for Computer Scientists and Mathematicians*, Second Edition, Prentice-Hall, Englewood Cliffs, New Jersey, 1986.

[MUDG91] Mudge, T.N., R.B. Brown, W.P. Birmingham, J.A. Dykstra, A.I. Kayssi, R.J. Lomax, O.A. Olukotun, K.A. Sakallah, and R.A. Milano, "The Design of a Microsupercomputer," *IEEE Computer*, Vol. 24, No. 1, pp. 57-64, January 1991.

[NELS86] Nelson, V.P., and B.D. Carroll, *Tutorial: Fault-Tolerant Computing*, IEEE Computer Society Press, Washington, DC, 1986.

[NELS90] Nelson, V.P., "Fault-Tolerant Computing: Fundamental Concepts," *IEEE Computer*, Vol. 23, No. 7, July 1990.

[NEUM56] von Neumann, J., "Probabilistic Logic and the Synthesis of Reliable Organisms," in *Automata Studies*, C.E. Shannon and J. McCarthy, Eds., Princeton University Press, Princeton, N.J., pp. 43-98, 1956.

[OPPE75] Oppenheim, A.V. and R.W. Schafer, *Digital Signal Processing*, Prentice-Hall, Englewood Cliffs, New Jersey, 1975.

[PATT85] Patterson, D.A., "Reduced Instruction Set Computers," *Communications of the ACM*, Vol. 28, No. 1, January 1985.

[PHIL84] Phillips, C. and H. Nagle, *Digital Control System Analysis and Design*, Prentice-Hall, Englewood Cliffs, New Jersey, 1984.

[PIER65] Pierce, W.H., *Failure-Tolerant Computer Design*, Academic Press, New York, NY, 1965.

[PRAD74] Pradhan, D.K. and S.M. Reddy, "Design of Two-Level Fault-Tolerant Networks," *IEEE Transactions on Computers*, Vol. C-23, No. 1, pp. 41-48, January 1974.

[PRAD83] Pradhan, D.K., "Fault-Tolerant Architectures for Multiprocessors and VLSI Systems," *Digest of the 13th Annual International*

Symposium on Fault-Tolerant Computing, Milan, Italy, pp. 436-441, June 1983.

[PRAD85a] Pradhan, D.K., "Fault-Tolerant Multiprocessor Link and Bus Architectures," *IEEE Transactions on Computers*, Vol. C-34, No. 1, pp. 33-45, January 1985.

[PRAD85b] Pradhan, D.K., "Dynamically Restructurable Fault-Tolerant Processor Network Architecture," *IEEE Transactions on Computers*, Vol. C-34, No. 5, pp. 434-447, May 1985.

[PRAD86a] Pradhan, D.K., editor, *Fault-Tolerant Computing: Theory and Techniques*, Volumes 1 and 2, Prentice-Hall, Englewood Cliffs, New Jersey, 1986.

[PRAD86b] Pradhan, D.K., Correction to "Fault-Tolerant Multiprocessor Link and Bus Architectures," *IEEE Transactions on Computers*, p. 94, Vol. C-35, No. 1, January 1986.

[RAMA90] Ramanathan, P., K.G. Shin, and R.W. Butler, "Fault-Tolerant Clock Synchronization in Distributed Systems," *IEEE Computer*, Vol. 23, No. 10, pp. 33-42, October 1990.

[RAND75] Randall, B., "System Structure for Software Fault Tolerance," *IEEE Transactions on Software Engineering*, Vol. SE-1, No. 1, pp. 220-232, June 1975.

[RAO89] Rao, T.R.N. and E. Fujiwara, *Error-Control Coding for Computer Systems*, Prentice-Hall, Englewood Cliffs, New Jersey, 1989.

[RENN80] Rennels, D.A., "Distributed Fault-Tolerant Computer Systems," *IEEE Computer*, Vol. 13, No. 3, pp. 55-64, March 1980.

[RENN84] Rennels, D.A., "Fault-Tolerant Computing - Concepts and Examples," *IEEE Transactions on Computers*, Vol. C-33, No. 12, pp. 1116-1129, December 1984.

[RUDO67] Rudolph, L.D., "A Class of Majority Logic Decodable Codes," *IEEE Transactions on Information Theory*, Vol. IT-13, No. 2, pp. 305-307, April 1967.

[SERL84] Serlin, O., "Fault-Tolerant Systems in Commercial Applications," *IEEE Computer*, Vol. 17, No. 8, pp. 19-30, August 1984.

[SHIN87] Shin, K.G. and P. Ramanathan, "Clock Synchronization of a Large Multiprocessor System in the Presence of Malicious Faults," *IEEE Transactions on Computers*, Vol. C-36, No. 1, pp. 2-12, January 1987.

[SHIN88] Shin, K.G. and P. Ramanathan, "Transmission Delays in Hardware Clock Synchronization," *IEEE Transactions on Computers*, Vol. C-37, No. 11, pp. 1465-1467, November 1988.

[SIEW78] Siewiorek, D.P., V. Kini, H. Mashburn, S.R. McConnel, and M.M. Tsao, "A Case Study of C.mmp, Cm*, and C.vmp: Part I-Experiences with Fault Tolerance in Multiprocessor Systems," *Proceedings of the IEEE*, Vol. 66, No. 10, pp. 1178-1199, 1978.

[SIEW82] Siewiorek, D.P. and R.S. Swarz, *The Theory and Practice of Reliable System Design*, Digital Press, Bedford, MA, 1982.

[SIEW84] Siewiorek, D.P., "Architecture of Fault-Tolerant Computers," *IEEE Computer*, Vol. 17, No. 8, August 1984.

[SIEW90] Siewiorek, D.P., "Fault Tolerance in Commercial Computers," *IEEE Computer*, Vol. 23, No. 7, July 1990.

[SIMO90] Simons, Jr., F.O. and A.D. George, "TMS320C30 vs. DSP96000: A Digital Filtering Comparison (Feasibility and Implementation Issues)," *Proceedings of the American Society for Engineering Education (ASEE) Annual Conference*, Toronto, Canada, June 1990.

[SING90] Singh, A.D., and S. Murugesan, "Fault Tolerant Systems," *IEEE Computer*, Vol. 23, No. 7, July 1990.

[SKLA76] Sklaroff, J.R., "Redundancy management technique for space shuttle computers," *IBM Journal of Research and Development*, Vol. 20, No. 1, pp. 20-28, January 1976.

[SMIT81] Smith III, T.B., "Fault-Tolerant Clocking System," *Digest of the 11th Annual International Symposium on Fault-Tolerant Computing*, Portland, Maine, pp. 262-264, June 1981.

[SMIT86a] Smith III, T.B., J.H. Lala, J. Goldberg, W.H. Kautz, P.M. Melliar-Smith, M.W. Green, K.N. Levitt, R.L. Schwartz, C.B. Weinstock, D.L. Palumbo, and R.W. Butler, *The Fault-Tolerant Multiprocessor Computer*, Noyes Publications, Park Ridge, New Jersey, 1986.

[SMIT86b] Smith III, T.B., "High Performance Fault-Tolerant Real-Time Computer Architecture," *Digest of the 16th Annual International Symposium on Fault-Tolerant Computing*, Vienna Austria, July 1986.

[SOHI88] Sohie, G.R.L. and K.L. Kloker, "A Digital Signal Processor with IEEE Floating-Point Arithmetic," *IEEE Micro*, Vol. 8, No. 6, pp. 49-67, December 1988.

[STAN88] Stankovic, J.A., "Misconceptions About Real-Time Computing: A Serious Problem for Next-Generation Systems," *IEEE Computer*, Vol. 21, No. 10, pp. 10-19, October 1988.

[STON82] Stone, H., *Microcomputer Interfacing*, Addison-Wesley, Reading, MA, 1982.

[SVOB89] Svobodova, L., "Attaining Resilience in Distributed Systems," from T. Anderson, editor, *Dependability of Resilient Computers*, BSP Professional Books, Oxford, England, pp. 98-124, 1989.

[TANE85] Tanenbaum, A.S. and R. Van Renesse, "Distributed Operating Systems," *ACM Computing Surveys*, Vol. 17, No. 4, December, pp. 419-470, 1985.

[VASA88] Vasanthavada, N. and P.N. Marinos, "Synchronization of Fault-Tolerant Clocks in the Presence of Malicious Failures," *IEEE Transactions on Computers*, Vol. C-37, No. 4, pp. 440-448, April 1988.

[WAKE75] Wakerly, J.F., "Transient Failures in Triple Modular Redundant Systems with Sequential Machines," *IEEE Transactions on Computers*, Vol. C-24, No. 5, pp. 570-573, May 1975.

[WENS78] Wensley, J.H., L. Lamport, J. Goldberg, M.W. Green, K.N. Levitt, P.M. Melliar-Smith, R.E. Shostak, and C.B. Weinstock, "SIFT: Design and Analysis of a Fault-Tolerant Computer for Aircraft Control," *Proceedings of the IEEE*, Vol. 66, No. 10, pp. 1240-1255, October 1978.

[WENS83] Wensley, J.H., "An Operating System for a TMR Fault-Tolerant System," *Digest of the 13th Annual International Symposium on Fault-Tolerant Computing*, Milano, Italy, pp. 452-455, June 28-30, 1983.

[WENS85] Wensley, J.H., "August Systems industrial control computers," from T. Anderson, T., editor, *Resilient Computing Systems*,

Collins Professional and Technical Books, London, England, pp. 232-246, 1985.

[WENS85] Wilson, D., "The STRATUS computer system," from T. Anderson, T., editor, *Resilient Computing Systems*, Collins Professional and Technical Books, London, England, pp. 197-231, 1985.

[WHIT85] Whitehouse, H.J., J.M. Speiser, and K. Bromley, "Signal Processing Applications of Concurrent Array Processor Technology," from S.Y. Kung, H.J. Whitehouse, and T. Kailath, editors, *VLSI and Modern Signal Processing*, Prentice-Hall, Englewood Cliffs, New Jersey, pp. 25-41, 1985.

[ZIMM80] Zimmerman, H., "OSI reference model - The ISO model of architecture for open systems interconnection," *IEEE Transactions on Communications*, COM-28, pp. 425-432, April 1980.

INDEX

WIDENER UNIVERSITY
WOLFGRAM
LIBRARY
CHESTER, PA.

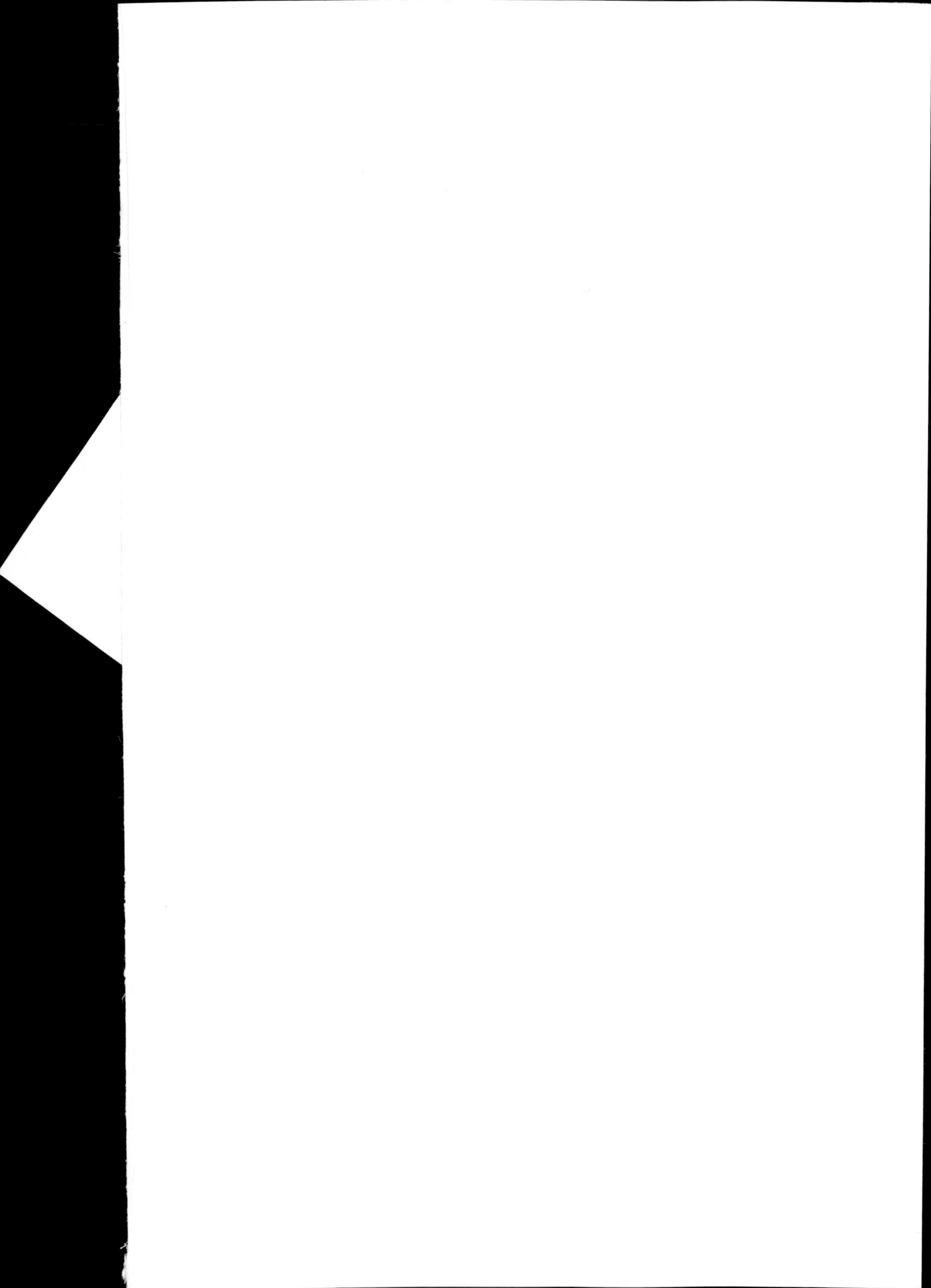

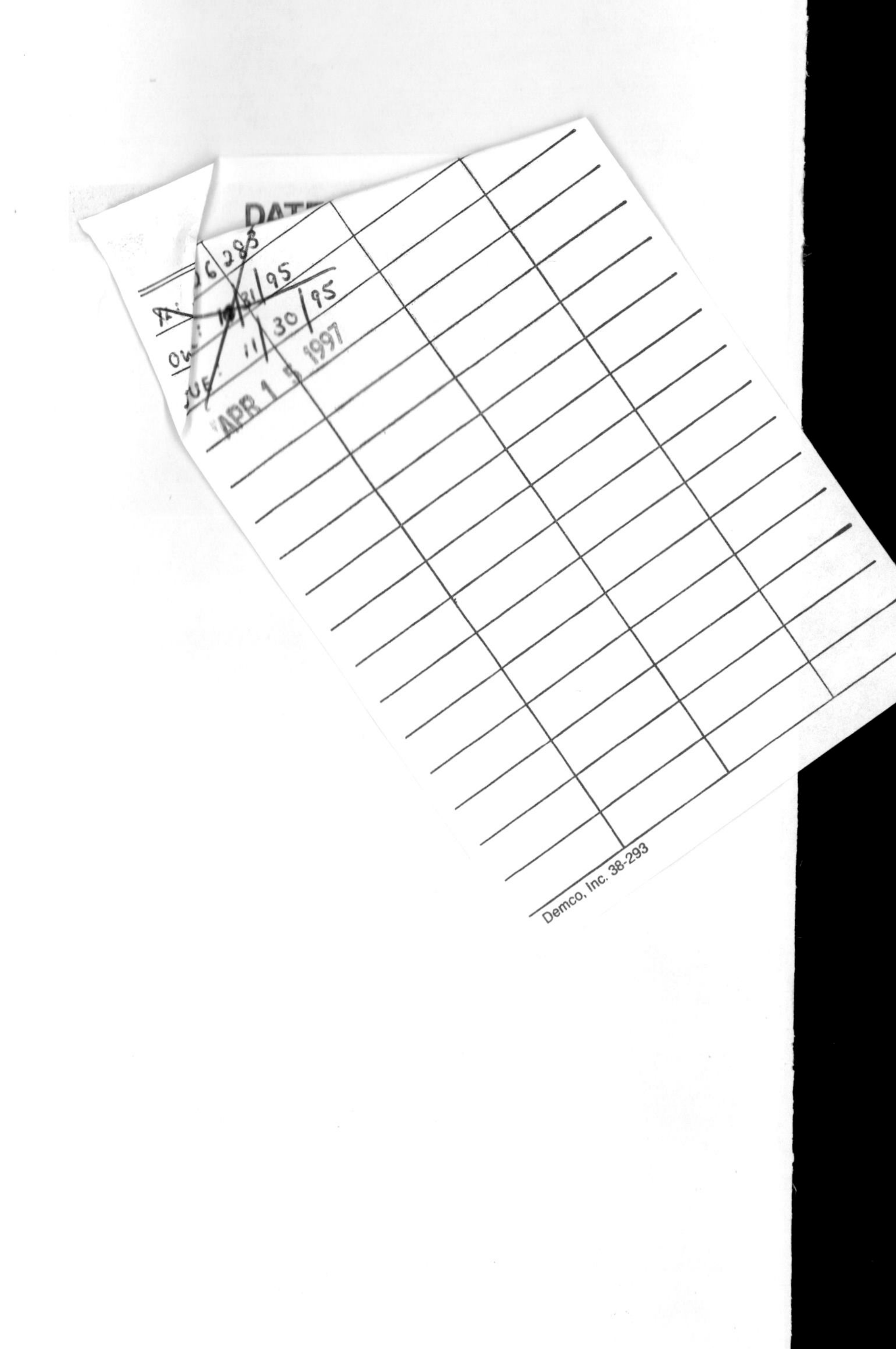
APR 1 5 1997
Demco, Inc. 38-293